SPACE SCIENCE, EXPLORATION AND POLICIES

Additional books in this series can be found on Nova's website under the Series tab.

Additional E-books in this series can be found on Nova's website under the E-books tab.

PULSARS

DISCOVERIES, FUNCTIO
AND FORMATION

PULSARS

DISCOVERIES, FUNCTIONS AND FORMATION

PETER A. TRAVELLE
EDITOR

Nova Science Publishers, Inc.
New York

Copyright © 2011 by Nova Science Publishers, Inc.

All rights reserved. No part of this book may be reproduced, stored in a retrieval system or transmitted in any form or by any means: electronic, electrostatic, magnetic, tape, mechanical photocopying, recording or otherwise without the written permission of the Publisher.

For permission to use material from this book please contact us:
Telephone 631-231-7269; Fax 631-231-8175
Web Site: http://www.novapublishers.com

NOTICE TO THE READER

The Publisher has taken reasonable care in the preparation of this book, but makes no expressed or implied warranty of any kind and assumes no responsibility for any errors or omissions. No liability is assumed for incidental or consequential damages in connection with or arising out of information contained in this book. The Publisher shall not be liable for any special, consequential, or exemplary damages resulting, in whole or in part, from the readers' use of, or reliance upon, this material. Any parts of this book based on government reports are so indicated and copyright is claimed for those parts to the extent applicable to compilations of such works.

Independent verification should be sought for any data, advice or recommendations contained in this book. In addition, no responsibility is assumed by the publisher for any injury and/or damage to persons or property arising from any methods, products, instructions, ideas or otherwise contained in this publication.

This publication is designed to provide accurate and authoritative information with regard to the subject matter covered herein. It is sold with the clear understanding that the Publisher is not engaged in rendering legal or any other professional services. If legal or any other expert assistance is required, the services of a competent person should be sought. FROM A DECLARATION OF PARTICIPANTS JOINTLY ADOPTED BY A COMMITTEE OF THE AMERICAN BAR ASSOCIATION AND A COMMITTEE OF PUBLISHERS.

Additional color graphics may be available in the e-book version of this book.

LIBRARY OF CONGRESS CATALOGING-IN-PUBLICATION DATA

Pulsars : discoveries, functions, and formation / [edited by] Peter A. Travelle.
    p. cm.
  Includes index.
  ISBN 978-1-61122-982-0 (hardcover)
  1. Pulsars. I. Travelle, Peter A.
  QB843.P8P85 2010
  523.8'874--dc22
                2010047000

Published by Nova Science Publishers, Inc. † *New York*

CONTENTS

PREFACE

A pulsar is a rapidly spinning neutron star that has a mechanism to beam light. This mechanism is only partially understood, but is connected with very strong magnetic fields spinning with the star. This book presents and discusses current research in the study of pulsars, including changes in the orbital periods of binary pulsars; pulsar distances and the electron distribution in the galaxy; magnetic field evolution through pulsar glitches; natal pulsar kicks; particle acceleration in pulsar outer magnetospheres and accretion-driven millisecond x-ray pulsars.

Pulsars appear to run around the galaxy (and its halo) with mean spatial (3-D) velocities of $\sim$450 kms^{-1}. Since most stars in our galaxy are observed to drift with mean velocities of $\sim$ 30 kms^{-1}, it has been suggested that such large velocities should be given to the proto-neutron star at its birth. The mechanism responsible for the impulses has not been properly identified yet, despite the interesting proposals that have been advanced. In Chapter 1 the author suggests that such a mechanism might be linked to the *radiation reaction force* of the powerful gravitational wave burst released from the gravity-driven r-mode instability underwent by the nascent rapidly rotating neutron star, remnant of a given supernova explosion. Since r-modes are associated with axial perturbations of the star fluid, then they can impart a recoil velocity to the nascent pulsar to launching it into a galactic or halo orbit. The resulting kick would be preferentially along the spin axis, a conclusion that seems to have been confirmed by observations of the Crab and Vela pulsars. Because the results presented are obtained by using Einstein's general relativity theory, and are supported by numerical simulations of both the non-linear evolution of r-modes of rapidly spinning neutron stars and effects of gravitational waves impinging compact stars, it turns out that the overwhelming population of fast moving pulsars cataloged, compared to the tiny one of neutron star-neutron star binaries, would turn these objects the most compelling evidence for the existence of gravitational waves. This result would correspondingly point, once again, towards Einstein's theory of gravitation as the most correct one.

In Chapter 2 the phenomena which yield to changes of the orbital periods of any binary systems are considered, and compared with the observed values for some binary pulsars. In the introduction the known formula caused by the gravitational radiation according to the general relativity, and the corrections caused by kinematic influence are presented. The method of evaluation of the decay of the orbital period via the advance of the periastron rate and the time dilation/gravitational red shift amplitude γ, is also presented. In the sections 2 and 3 are presented some recent results about time dependent gravitational potential, which also cause change of the orbital period on any binaries. It is assumed that the gravitational potential in the universe changes linearly, or almost linearly, with the time via $\Delta V = -c^2 H \Delta t$. This change is determined such that it is compatible with the Hubble red shift. It causes some deformations of the trajectories of the planetary orbits as well as the orbits of any binaries, such that the trajectories are not axially symmetric. The basic influence is the increasing the orbital period of any binary system, such that $\dot{P_b} = P_b H / 3$. In section 4 are considered the changes of the orbital periods of four pulsars according to the gravitational radiation together with the influence from the universe, and it is compared with the observed change of the orbital periods. In case of the Hulse-Taylor pulsar B1913+16 the dominant role has the gravitational radiation and it confirms very well; in case of the pulsar B1885+09 the dominant role has the influence from the gravitational potential and it confirms, and in case of the pulsar B1534+12 simultaneously confirm both the gravitational radiation and the gravitational potential in the universe. In the section 5 are considered some additional arguments about the changes and observations of the distances, times and velocities, caused via the gravitational potential in the universe. Using these arguments, in section 6 is discussed the correlation between the change of the distance Earth-Moon, the increasing of the lunar orbital period and the increasing of the average length of day.

In Chapter 3, the authors present a phenomenological approach to determine pulsar distances and the average electron distribution in the Galaxy. The authors present the pulsar distances as a function of dispersion measure represented as graphs, not as closed mathematical formulae. The method is analogous to the widely used method which utilizes the optical absorption (A_V) to determine distances of stars in different directions. To have reliable dispersion measure - pulsar distance relations, the authors have used several natural requirements and distances of pulsar-calibrators. Having obtained trustworthy distances, the authors have constructed dispersion measure - distance relations for pulsars for 48 different regions in the Galaxy. Using the dispersion measure - distance graphs in this work, the distance to any pulsar can be determined if its dispersion measure value is known.

As explained in Chapter 4, glitches are thought to be a kind of common phenomena in pulsars these days. From the observations, the authors know that after each glitch, there is often a permanent increase in the pulsar's spin-down rate (period derivative). Therefore, a pulsar's present spin-down rate may be much higher than its initial value and the estimated characteristic age of a pulsar based on its present spin-down rate and period may be shorter than its true age. At the same time, the permanent increase of its spin-down rate implies that the pulsar's surface magnetic field is increased after each glitch. Consequently, after many glitches, some radio pulsars may evolve into magnetars, i.e., strongly magnetized and slowly rotating neutron stars.

Pulsars are rapidly rotating neutron stars and are the outcome of the collapse of the core of a massive star with a mass of the order of or larger than eight solar masses. This process releases a huge gravitational energy of about 10^{53} erg, mainly in the form of neutrinos. During the collapse the density increases, and so does the magnetic field due to the trapping of the flux lines of the progenitor star by the high conductivity plasma. When the density reaches a value of around 10^{12} g cm^{-3} neutrinos become trapped within the protoneutron star and a neutrinosphere, characterized inside by a diffusive transport of neutrinos and outside by a free streaming of neutrinos, is formed and lasts for a few seconds. In Chapter 5 the authors focus on the structure of the neutrinosphere, the resonant flavor conversion that can happen in its interior, and the neutrino flux anisotropies induced by this phenomena in the presence of a strong magnetic field. The authors present a detailed discussion in the context of the spherical Eddington model, which provides a simple but reasonable description of a static neutrino atmosphere, locally homogenous and isotropic. Energy and momentum are transported by neutrinos and antineutrinos flowing through an ideal gas of nonrelativistic, nondegenerate nucleons and relativistic, degenerate electrons and positrons. The authors examine the details of the asymmetric neutrino emission driven by active-sterile neutrino oscillations in the magnetized protoneutron star, and the possibility for this mechanism to explain the intrinsic large velocities of pulsars respect to nearby stars and associated supernova remnants.

The first part of Chapter 6 gives the introductory words on pulsars. Then, in the second part of our text concerning the gravitational and electromagnetic radiation of pulsars, the authors have derived the total quantum energy loss of the binary. The energy loss is caused by the emission of gravitons during the motion of the two binary bodies around each other under their gravitational interaction. The energy-loss formula of the production of gravitons is derived here in the source theory formulation of gravity. Because the general relativity does not necessarily contain the last valid words to be written about the nature of gravity and it is not, of course, a quantum theory, it cannot give the answer on the production of gravitons and the quantum energy loss, respectively. So, this is the treatise that discusses the quantum energy loss caused by the production of gravitons by the binary system.

In the third part of this chapter, the quantum energy loss formula involving radiative corrections is derived also in the framework of the source theory. The general relativity does not necessarily contain the method how to express the quantum effects together with the radiative corrections by the geometrical language. So it cannot give the answer on the production of gravitons and on the graviton propagator with radiative corrections. So, the present text discusses the quantum energy loss caused by the production of gravitons and by the radiative corrections in the graviton propagator in case of the motion of binary.

At the present time the only classical radiation of gravity was confirmed. The production of gravitons is not involved in the Einstein theory. However, the idea, that radiative corrections can have macroscopic consequences is logically clear. The authors present an idea that radiative corrections effects might play a role in the physics of cosmological scale. The situation in the gravity problems with radiative corrections is similar to the situation in QED in time many years ago when the QED radiative corrections were theoretically predicted and then experimentally confirmed for instance in case of he Lamb shift, or, of the anomalous magnetic moment of electron.

Astrophysics is, therefore, in crucial position in proving the influence of radiative corrections on the dynamics in the cosmic space. The authors hope that the further astrophysical observations will confirm the quantum version of the energy loss of the binary with graviton propagator with radiative corrections.

In the fourth part of the chapter, the authors have derived the power spectrum formula of the synchrotron radiation generated by the electron and positron moving at the opposite angular velocities in homogeneous magnetic field. The authors have used the Schwinger version of quantum field theory, for its simplicity. It is surprising that the spectrum depends periodically on radiation frequency ω and time which means that the system composed from electron, positron and magnetic field behaves as a pulsating system. While such pulsar can be represented by a terrestrial experimental arrangement it is possible to consider also the cosmological analogue.

To our knowledge, our result is not involved in the classical monographs on the electromagnetic theory and at the same time it was not still studied by the accelerator experts investigating the synchrotron radiation of bunches. This effect was not described in textbooks on classical electromagnetic field and on the synchrotron radiation. The authors hope that sooner or later this effect will be verified by the accelerator physicists.

The treatise involves modification of an older article, in which the spectral formula of the gravitational radiation was derived, author's CERN preprints and the author preprints and articles on the electromagnetic pulsars.

In the last 15 years our knowledge on the high-energy emission from rotation-powered pulsars has considerably improved. The seven known gamma-ray pulsars provide us with essential information on the properties of the particle accelerator — electro-static potential drop along the magnetic field lines. Although the accelerator theory has been studied for over 30 years, the origin of the pulsed gamma-rays is an unsettled question. In Chapter 7, the authors review accelerator theories, focusing on the electrodynamics of outer-magnetospheric gap models. Then the authors present a modern accelerator theory, in which the Poisson equation for the electrostatic potential is solved self-consistently with the Boltzmann equations for electrons, positrons, and gamma-rays.

In Chapter 8 the author presents an overview of our current observational knowledge of the six known accretion-driven millisecond X-ray pulsars. A prominent place in this review is given to SAX J1808.4--3658; it was the first such system discovered and currently four outbursts have been observed from this source, three of which have been studied in detail using the *Rossi X-ray Timing Explorer* satellite. This makes SAX J1808.4--3658 the best studied example of an accretion-driven millisecond pulsar. Its most recent outburst in October 2002 is of particular interest because of the discovery of two simultaneous kilohertz quasi-periodic oscillations and nearly coherent oscillations during type-I X-ray bursts. This is the first (and so far only) time that such phenomena are observed in a system for which the neutron star spin frequency is exactly known. The other five systems were discovered within the last three years (with IGR J00291+5934 only discovered in December 2004) and only limited results have been published.

In: Pulsars: Discoveries, Functions and Formation
Editor: Peter A. Travelle, pp. 1-16

ISBN 978-1-61122-982-0
© 2011 Nova Science Publishers, Inc.

Chapter 1

NATAL PULSAR KICKS
FROM BACK REACTION
OF GRAVITATIONAL WAVES

Herman J. Mosquera Cuesta[1,2]*
[1]Instituto de Cosmologia, Relatividade e Astrofísica (ICRA-BR),
Centro Brasileiro de Pesquisas Físicas, Rua Dr. Xavier Sigaud 150,
Cep 22290-180, Urca, Rio de Janeiro, RJ, Brazil
[2]Abdus Salam International Centre for Theoretical Physics,
Strada Costiera 11, Miramare 34014, Trieste, Italy

Abstract

Pulsars appear to run around the galaxy (and its halo) with mean spatial (3-D) velocities of $\sim 450\,\mathrm{kms}^{-1}$. Since most stars in our galaxy are observed to drift with mean velocities of $\sim 30\,\mathrm{kms}^{-1}$, it has been suggested that such large velocities should be given to the proto-neutron star at its birth. The mechanism responsible for the impulses has not been properly identified yet, despite the interesting proposals that have been advanced. Here I suggest that such a mechanism might be linked to the *radiation reaction force* of the powerful gravitational wave burst released from the gravity-driven r-mode instability underwent by the nascent rapidly rotating neutron star, remnant of a given supernova explosion. Since r-modes are associated with axial perturbations of the star fluid, then they can impart a recoil velocity to the nascent pulsar to launching it into a galactic or halo orbit. The resulting kick would be preferentially along the spin axis, a conclusion that seems to have been confirmed by observations of the Crab and Vela pulsars. Because the results presented are obtained by using Einstein's general relativity theory, and are suported by numerical simulations of both the non-linear evolution of r-modes of rapidly spinning neutron stars and effects of gravitational waves impinging compact stars, it turns out that the overwhelming population of fast moving pulsars cataloged, compared to the tiny one of neutron star-neutron star binaries, would turn these objects the most compelling evidence for the existence of gravitational waves. This result would correspondingly point, once again, towards Einstein's theory of gravitation as the most correct one.

*E-mail address: hermanjc@cbpf.br

1. Introduction

Gravitational waves (GWs) are ripples in the fabric of spacetime generated by cataclysmic events such as supernova (SN) core-collapse and explosion, the non-symmetric evolution of its compact remnant, and coalescences of binary neutron star (NS) and/or black holes (BHs), among other astrophysical sources. Its direct detection still awaits for firm confirmation. However, indirect evidences for their existence come from the observations by Taylor and Hulse of the dynamical evolution of the *binary pulsar* PSR 1913+16 (Taylor 1993,2003). During a SN core-collapse, GWs are thought to be produced by either the hydrodynamical evolution of the protoneutron star (PNS) or the neutrino outflow after the core bounce (Burrows et al. 1995; Burrows & Hayes 1996; Janka & Müller 1996; Müller & Janka 1997; Mosquera Cuesta 2000,2002; Loveridge 2004; Mosquera Cuesta & Fiuza 2004).

Natal pulsar kicks (recoil velocities), as inferred from all-sky surveys, range over 100-1500 km s^{-1} (Lorimer et al. 1993; Lyne & Lorimer 1994; Kaspi et al. 1996; Cordes & Chernoff 1998), while their rotation periods pile up near 0.5 seconds (Taylor 1993; Lorimer et al. 1993; Camilo et al. 1996). These properties cannot be explained as inherited of the late stellar evolutionary stages of their progenitor supergiant stars (SGSs), since the cores of these SGSs rotate so slowly as to leave remnant pulsars with millisecond spin periods and high velocities. As a consequence, the origin of the pulsar recoil velocity should unavoidably be associated to the PNS formation process itself.

Here I skip from those *conventional mechanisms* to kick the PNS, as either asymmetric hydrodynamic explosions or neutrino convectively advected minijets (Janka & Müller 1994; Cowsik 1998; Spruit & Phinney 1998). It is suggested instead that due to its *multipolar nature* the *radiation-reaction force* (RRF) of the *gravitational wave (long) burst* from *r*-modes, a kind of axial perturbation, see below, is a likely mechanism to kick the pulsars. This is a very different argumentation, that goes in the lines of a separate work by Mosquera Cuesta (2002); in which large amounts of GWs are produced from neutrino oscillations during the SN core bounce. Thence, just after at the core bounce of a SN collapse recoil velocities and spins may be imparted to the PNS as kicks at birth via GWs. As the GWs go away from the NS the RRF induced may apply several (pair-folded) thrusts to the nascent pulsar. This effect could explain its rotation and translational motion as evidenced by pulsar surveys (Lorimer et al. 1993; Lyne & Lorimer 1994; Cordes & Chernoff 1998).

However, we stress in passing that, as noted by some workers in the field, the resulting kick in this mechanism would be preferentially along the rotation axis unless an additional mechanism (magnetic field stresses, hydrodynamic jets, or neutrino emission) is found able to act simultaneously with the *r*-modes to drive the mode off-centered with respect to the spinning axis. This combined action should result in no azimuthal averaging of the radiation reaction force after thousands of runs of the young neutron star. This may produce kicks uniformly distributed, and in passing it may bring the whole theory here proposed in full agreement with pulsar observations. This key feature is reviewed below. In the meantime, it should be emphasized that the issue deserves a further investigation.

1.1. Astrophysical Hints at GWs-driven Spinning NSs

That GWs emission could indeed be responsible for the pulsar runaway velocities was conjectured a few years ago by Nazin & Postnov (1997). An inference from the proper motion
of pulsars suggests that in order for GWs to push out the nascent NS, an energy in GWs:
$E_{surveys}^{Pulsar}(\text{GWs}) \sim 4 \times 10^{-6} M_\odot c^2$ should be released during the early evolutionary phases
(Nazin & Postnov 1997). As presented here, gravitational radiation reaction from the r-
mode could do the work of imparting recoil momentum when the NS is born.

From a theoretical point of view, we note that numerical simulations of GWs generation
during a SN core-collapse suggest that, in the most optimistic case, only about a fraction of
$\Delta E_{\text{GWs}} \sim [10^{-10} - 10^{-13}] M_\odot c^2$ of the total binding energy of the PNS is released as GWs
from the NS fluid hydrodynamics (Burrows & Hayes 1996; Müller & Janka 1997). It has
been shown that such a tiny energy cannot by itself do the work of kicking the star (Janka
& Müller 1994; Lai, Chernoff & Cordes 2000, Lai 2000). However, as we demonstrate in
this work, the effect of the huge GWs *luminosity* radiated as the r-mode instability develops
in the NS rotating fluid may manifest itself through this RRF in the form of pulsar kicks.
This instability shows up as an axial perturbation of the star fluid, and is characterized by
an extremely short damping time, as shown below.

2. Pulsar Surveys and Implications for Viable Kick Mechanisms

We will start this section by reviewing the status of the pulsar astrophysics by the late
1990's, and then we make the connection with our current understanding of all the outstanding features of this fast developing research field.

In a very interesting review on the observational status of the astrophysics of pulsars
circa 1999, Deshpande, Ramachandran & Radakrishnan (1999, DRR99) arrived, based
mainly in radio observations, to some crucial conclusions concerning the direction and
magnitude of the proper motion, projected direction of the magnetic axis, magnitude of the
magnetic field, direction of the rotation axis and initial spin periods (which are not properly
inferred from observations) of pulsars. These authors also showed that no significant correlation exists between the magnetic field strength and the magnitude of the spatial velocities
of pulsars, or between the projected directions of the rotation axis (and/or the magnetic field
axis) and the direction of the proper motion vector.

Such conclusions have fundamental implications for the mechanisms proposed to trigger asymmetric SNe and recoil velocities, as by the time of the work by DRR99 the observations did not support any mechanism producing net kick velocities parallel to the rotational
axis. (Only after the advent of CHANDRA the view on these issues changed radically,
see below). Consequently, the DRR99 study ruled out momentum impulses of any duration along the spinning axis, and any long duration impulses (compared to the spin period)
along any one fixed axis, as the magnetic vector. In addition, their numerical simulations
and confrontation with observations altogether led them to conclude that *single impulses*
are also *ruled out*, but not *two or more impulses* of relatively short duration (Deshpande,
Ramachandran & Radakrishnan 1999). Moreover, in order to avoid significant azimuthal
averaging of the impulse radial component, the momentum transfer event should be very
short-lived. Those conclusions apply to all the proposed mechanisms to kick the pulsars:

Spruit & Phinney (1998), Cowsik (1998) and Lai, Chernoff & Cordes (2000), etc., thought to be the most prospective ones. As quoted above, the mechanism introduced in this paper leads to kicks aligned with the rotation axis, but certainly competing effects may help to force it to be off-centered with respect to that spatial direction.

Accordingly, all the mechanisms driving pulsar kicks from neutrino oscillations, such as those by Kusenko & Segrè (1997,1998); Akhmedov, Sciama & Lanza (1998); Grasso et al. (1998); Nardi & Zuluaga (2000); Mosquera Cuesta (2000);, as well as the Harrison-Tademaru (1975) *rocket-like mechanism* (Lai 2000), are critically questioned since for all of them to properly work, a dipolar magnetic field configuration pervading the PNS core is a crucial assumption, and only one thrust is unavoidable. Conversely, only those scenarios in which impulses triggered in very short timescales and almost simultaneously applied to the PNS (Spruit & Phinney 1998, Cowsik 1998) are in principle capable of accelerating and make it to spin the nascent NS as observed in pulsar surveys (Lyne & Lorimer 1994; Kaspi 1996). Below we demonstrate that those constraints can be consistently satisfied by the mechanism proposed in this paper involving GWs. As it will be discussed later on, the possibility of most pulsars to be born with a few millisecond periods is straightforwardly realized within our GWs-RRF-driven pulsar kick mechanism.

I stress, however, that the above view by DRR99 changed drammatically after the observations of the Crab and Vela pulsars with the CHANDRA X-ray telescope (Weisskopf 2000). The spectacular, high resolution images of the Crab pulsar clearly reveals such an association. This confirmed association of the Crab pulsar proper motion with the spinning axis definitely rules out the conclusions put forward by DRR99, and opens a new perspective in the pulsar physics that seems to have room for the *r*-mode instability to play a fundamental role. We explore this possibility next.

3. Gravitational Waves in Einstein's General Relativity

In general relativity, the Einstein's theory of gravitation, the dynamics of a pulsating compact star is described according to the field equations

$$R_{\mu\nu} - \frac{1}{2}g_{\mu\nu}R = \kappa T_{\mu\nu} \tag{1}$$

where $R_{\mu\nu}$, R, $g_{\mu\nu}$ and $T_{\mu\nu}$, are correspondently the curvature tensor, curvature scalar and metric tensor of spacetime, and energy-momentum tensor of the matter fields constituting the body. κ is a constant. When linearized, for taking into consideration small perturbations ($h << 1$) of the Minkowski spacetime, the metric can be written as

$$g_{\mu\nu} = n_{\mu\nu} + h_{\mu\nu}. \tag{2}$$

These equations lead to the gravitational waveform, the time evolution of the GW spacetime strain

$$h_{ij}^{TT} = h_+ e_{ij}^+ + h_\times e_{ij}^\times, \tag{3}$$

which is a traceless, transverse (to the propagation direction) and symmetric spatial tensor, with polarization tensor components as $e_{xx}^+ = e_{yy}^+ = 1$ and $e_{xy}^\times = e_{yx}^\times = 1$, with any other combination of indexes being null. The components $h_+ e_{ij}^+$ and $h_\times e_{ij}^\times$ appearing in Eq.(3) are denominated the polarization modes of the gravitational wave. We will come back to this crucial property when discussing the pulsar kick mechanism from the GWs bursts at core bounce. This h_{ij}^{TT} is analogous to the vector potential A in electrodynamics in the Lorentz gauge. There, by using the Maxwell's equations, the following vacuum relations can be written

$$A_0 = 0, \qquad A_{i,i} = 0, \qquad \partial^i \partial_j A_i = 0. \tag{4}$$

The correponding expressions for the case of the gravitational waves reads

$$h_{0j}^{TT} = 0, \qquad h_{ij,j}^{TT} = 0, \tag{5}$$

and the Einstein's field equation (the GW equation) ina vacuum reads

$$\partial^i \partial_j h_{ij}^{TT} = 0. \tag{6}$$

To get this set of equations use has been done of the *traceless* property of the metric tensor.

When a GW passes through a compact body (the NS, in the case) the PNS oscillates in a way described by the usual equations of motion, but the oscillations themselves are driven by the GW force

$$F_i = \frac{1}{2} m \ddot{h}_{ij}^{TT} \delta_j \tag{7}$$

acting on each element of the mass m. Therefore, the GW carries energy and momentum, i. e., it does some work. Due to its quadrupolar nature in general relativity, the energy and momentum densities cannot be localized at a given point but distributed over a few wavelengths (λ_{GW}) around the star. We shall see next that for the case of nascent neutron star this λ_{GW} will correspond roughly to the impact parameter for the GWs quadrupole-shaped force, conjectured here to be the trigger of the PNS kick.

The stress-energy tensor for GWs propagating in the z-direction has non-null components

$$T^{00} = \frac{T^{0z}}{c} = \frac{T^{zz}}{c^2} = \frac{c^2}{16\pi G} < (\dot{h}_+)^2 + (\dot{h}_\times)^2 > \tag{8}$$

where the brackets denote averaging over several wavelengths. Here T_{00} is the energy density, T^{z0} the energy flux, and T^{zz} the momentum flux released during the first seconds after the supernova core bounce.

4. Gravitational Waves from r-modes of Rapidly Rotating NSs

4.1. Supernova Physics and Just-Born NSs

In a very simplified picture, the mechanism of a SN explosion invokes the transfer of energy from the stellar core (the progenitor of the PNS remnant) and the SGS mantle that is blown away. The source of the huge energy transfer required to eject the envelope is assumed to

---

be a neutrino flash from the core deleptonization process. Since neutron matter cannot be packed beyond the nuclear density the star core bounces. During the next first milliseconds the central core density gets so high that no radiation nor even neutrinos can escape. They get themselves frozen-in. The whole PNS dynamics is governed then by gravity alone. Over that timescale, the sound wave generated at bounce builds up itself into a shock wave which pushes out the infalling matter in the envelope, while the neutrino momentum transfer drives the final explosion. For a progenitor star of intermediate mass, the remnant of the SN explosion is supposed to be a young, hot and rapidly spinning NS that can undergo the GWs-driven Chandrasekhar-Friedman-Schutz (CFS) instability during its early evolution.

4.2. The r-modes Instability

The CFS instability, in particular the r-mode instability, can develop in rapidly rotating NSs, as discovery by Andersson (1998) and Friedman and Morsink (1998). This instability sets in when back-reaction of gravitational radiation from the spinning star's fluid amplifies an oscillation mode that propagates forward with respect to an inertial frame, but backward with respect to the rotating fluid. It was pointed out firstly by Andersson (1998) that the r-modes of rotating NSs can be driven unstable due to back-reaction of GWs being emitted from the just born hot star (see recent reviews by Andersson & Kokkotas 2000; Lindblom 2001, and the numerical study performed by Yoshida, et al. 2000; Yoshida et al. 2004). Initial estimates of the instability timescale suggested that the GWs driving force is more stronger than the stabilizing effects from the NS matter (bulk and shear) viscosity (Lindblom, Owen & Morsink 1998).

Several other dissipative mechanisms, however, have been shown to strongly act against the growth (to the maximum amplitude) of the r-mode instability. A short list of those include: a) the formation of a solid crust on the NS (Bildsten & Ushomirsky 2000), b) the inclusion of magnetic fields (Rezzolla, et al. 2000), and c) the non-linear effects of hyperon bulk viscosity (Jones 2001a, Owen & Lindblom 2001). Although in several cases those additional effects are shown to lead, over long time scales, to a full suppression of the instability development, it by no means implies that the instability cannot play a rôle at all in the astrophysics of any young NSs (Owen & Lindblom 2001). Nowadays, the issue seems to be controversial: several groups support the idea of the development of the instability, whereas others contend it.

Despite of those hints just discussed, whose effects become manifest after long timescales compared to the mode growth times, in this work we propose that the r-mode instability can develop, at least during the relatively short timescale (Watts & Andersson 2002):

$$\Delta T_{r-\text{mode}}^{\text{growth}} \simeq -47\text{s} \left(\frac{1.4M_\odot}{M_{\text{NS}}}\right) \left[\frac{10\text{km}}{R_{\text{NS}}}\right]^4 \left(\frac{P}{10^{-3}\text{s}}\right)^6. \tag{9}$$

for a NS spinning with period $P_0 = (1\text{-}2)$ ms, so as to significantly effect the PNS dynamics via kicks at birth before the mode being effectively damped. (The "-" sign stands for the instability appearance). Because the early non-linear evolution of the r-modes of NSs is a moderately long-lived phase (see Lindblom, Tohline & Vallisneri (2001), LTV2001), during which the maximum of the GWs emission is reached after releasing an energy

$$\Delta E_{r-\text{mode}}^{\text{GWs}} \sim 50\% E_{\text{total}}^{\text{Spin}}, \tag{10}$$

we do expect the radiation reaction to be able to induce a recoil velocity on the nascent NS so as to launch it into a galaxy disk or halo orbit, as observed in pulsar surveys (Lorimer, et al. 1993; Lyne & Lorimer 1994; Cordes & Chernoff 1998).

Because of the NS emission of GWs the $l = m = 2$ r-mode, the one supposed to drive the more energetic growth of the instability, is expected to grow on a timescale $\Delta T_{r-\text{mode}}^{\text{growth}} \sim 47$s. This result indicates that a canonical NS spinning with period of ~ 1 ms will have its r-modes effectively driven unstable, as far as the shear and bulk viscosity damping effects are not so effective in suppressing the r-mode growth. The corresponding damping times of these competitive effects follow, respectively, as:

$$\Delta T_{\text{shear}}^{\text{growth}} \simeq 7.4 \times 10^7 \text{s} \left(\frac{0.1}{\alpha} \right)^{-5/3} \left(\frac{1.4 M_\odot}{M_{\text{NS}}} \right)^{5/9} \left[\frac{10 \text{km}}{R_{\text{NS}}} \right]^{-11/3} \left(\frac{10^9 \text{K}}{T_{\text{NS}}} \right)^{-5/3}, \tag{11}$$

and

$$\Delta T_{\text{bulk}}^{\text{growth}} \simeq 7.9 \text{ s} \left(\frac{1.4 M_\odot}{M_{\text{NS}}} \right)^{-2} \left[\frac{10 \text{km}}{R_{\text{NS}}} \right]^4 \left(\frac{P}{10^{-3} \text{s}} \right)^2 \left(\frac{10^9 \text{K}}{T_{\text{NS}}} \right)^2 \left(\frac{100 \text{MeV}}{m_{part.}} \right)^4. \tag{12}$$

Hence, a long and powerful GWs burst will be emitted over the r-mode growth time. Note, in passing, that the actual saturation time: $\Delta T_{r-\text{mode}}^{\text{growth}} \sim [30-40]$s (the one we use hereafter to compute the GWs luminosity), is more longer than the r-mode "growth time": $\Delta T \sim 26 P_0 = 10^{-2}$ s. LTV2001 succeeded in getting the r-modes non-linearly saturated, that is, the maximum value of the mode dimensionless amplitude (α in their usage) being reached after artificially enlarging the magnitude of the GWs-RRF in order to save computing time comsumption. Nonetheless, this procedure by no means comes to invalidate their modeling since in a realistic NS such saturate state can be reached provided no suppression effects appear from bulk and shear viscosities. Therefore, their results, in particular the total energy released during the process, still holds provided the only mode-damping mechanism is the breaking of surface shock waves (LTV2001), as we assume hereafter (see Yoshida et al. 2004).

Thus the long GWs burst released from these axial perturbations, and the mode's energy given away, may become, through their RRF, the impulse source required to impart the natal recoil velocity and spin to NSs at birth (Nazin & Postnov 1997; Mosquera Cuesta 2002; Mosquera Cuesta & fiuza 2004). Because the GWs-driven r-mode instability is excited as early as the PNS is formed, I suggest here that this GWs radiation reaction might be a fundamental mechanism responsible for the observed drift velocities and rotation periods of pulsars (Lorimer, et al. 1993; Lyne & Lorimer 1994; Cordes & Chernoff 1998).

5. Non-linear Evolution of r-modes and GWs Power

According to the non-linear follow-up of the r-modes in NSs by LTV2001, the energy and angular momentum taken away by GWs satisfy the relation

$$L_{\text{GWs}}^{r-\text{mode}} \equiv \frac{dE}{dt} = \frac{|\omega|}{2}\frac{dJ}{dt} = -\frac{128\pi G}{225c^7}|A|\omega^6|J_{22}|^2. \tag{13}$$

In the above equation the term $|J_{22}| = J_{22}e^{-i\psi(t)}$ (with $d\psi/dt = \omega$ the GWs angular frequency,) defines the time-dependent quadrupole current tensor

$$J_{22} = \int \rho r^2 \vec{v} \bullet \vec{Y}_{22}^{B\star} d^3x, \tag{14}$$

with $\vec{Y}_{22}^{B\star} = \hat{r} \times r\vec{\nabla}Y_{22}/\sqrt{6}$ as the magnetic-type vector spherical harmonic. In Eq.(13) the parameter $|A|$ measures the strength of the GWs back-reaction force, and is defined as: $|A| = \frac{32\sqrt{\pi}G}{45\sqrt{5}c^7}$, in Einstein's gravity. (It was the strength of this quantity that LTV2001 enlarged intensionally to save computing time in their numerical non-linear evolution of r-modes of NSs[25]). Thus Eq.(13) leads to a GWs luminosity

$$L_{\text{GW}}^{r-\text{mode}} \sim 3 \times 10^{51}\text{ergs}^{-1}\left(\frac{\Delta E_{\text{GWs}}^{r-\text{mode}}}{10^{-1}M_\odot c^2}\right)\left[\frac{47\text{s}}{\Delta T_{r-\text{mode}}^{\text{growth}}}\right], \tag{15}$$

where $\Delta T_{r-\text{mode}}^{\text{growth}} \sim 47\text{s}$ was used as the time interval for the maximum GWs emission during the saturation phase of the r-modes. Again, we have assumed that the mode amplitude can grow as large as the one numerically computed by LTV2001, which could be so whenever the unique mechanism to suppress it be the formation of breaking (hydro) shock waves at the NS surface.

By relating the flux of the outgoing gravitational waves to their luminosity, i. e.,

$$\frac{c^3}{16\pi G}\left|\frac{dh}{dt}\right|^2 = \frac{1}{4\pi D^2}L_{\text{GW}}, \tag{16}$$

one can obtain, after a time integration, the GWs waveform

$$h(t) = \frac{16}{15}\sqrt{\frac{2\pi}{5}}\left(\frac{\omega^2 G|J_{22}|}{c^5 D}\right)e^{i\psi(t)}, \tag{17}$$

where D is the distance to the pulsar. From this, the normalized GW strain amplitude follows as

$$h_{r-\text{mode}} \simeq 7.5 \times 10^{-21}\text{Hz}^{-1/2}\left[\frac{55\text{kpc}}{D}\right]$$

$$\times \left(\frac{980\text{Hz}}{f_{\text{GWs}}}\right)^2\left[\frac{|J_{22}|}{10^{54}\text{gcm}^3\text{s}^{-1}}\right]. \tag{18}$$

Here f_{GW} is the frequency of the GWs emitted in the process. A GW signal such as this from a future supernova in the galaxy could surely be observed by the initial LIGO, VIRGO, etc. interferometers. It could also be detected from SN events in the VIRGO cluster of galaxies by its Advanced Configuration, as shown in the Fig.4 of the paper by Owen & Lindblom (2001).

6. r-mode Radiation Reaction and Pulsar Kicks

A GW exerts a back reaction on a given source by virtue of carring away energy, angular and linear momentum. Whatever energy and momentum the wave takes away from a source will be reflected in this back reaction. Since axial (i.e., r-mode) perturbations can only drive pulsations of the PNS fluid through either a non-zero shear modulus or coupling to polar perturbations (in our case because the nascent PNS is hot and rapidly rotating), then the RRF of the r-mode GWs may apply the required thrusts to make it move and rotate at birth. Upon the need of some directional and asymmetric emission of the GWs to kick the PNS, one can in fact expect that not only the mass quadrupole tensor but higher multipoles, in particular the current quadrupole moment, to play an important rôle in kicking the pulsar due to its association to axial perturbations as the r-modes. Those contributions are computed next based upon this premise, together with the radiation emission beaming angle. A viable mechanism for an effective symmetry breaking along the pointing axis of the RRF direction is presented the appendix.

6.1. RRF of r-modes: Mass-Current Multipoles

The wave solution, Eq.(17) above, effectively include radiation-reaction effects relevant to the hydrodynamical description of a fluid of internal velocity v driven by pure GWs emission from r-modes. In a more general approach, the spacetime metric $g_{\mu\nu}$ is expanded in the parameter $(v/c)^{1/2}$ up to the post-Newtonian order $(v/c)^{7/2}$ (Blanchet 1997; Rezzolla et al. 1999), to take into account the rôle of mass and current multipoles. By following the Rezzolla et al. (1999) approach, the pure time-dependent mass and current quadrupole moments of the r-modes GWs RRF are written as (see Rezzolla et al. 1999, for details of the derivation and definitions)

$$F_i^{\frac{7}{2}\mathrm{PN}} = \bar{\rho}[-\partial_i(_9\alpha) - \partial_t(_8\beta_i) - \partial_j(_8\beta_i)v_j + \partial_i(_8\beta_j)v_j$$
$$- \partial_t(_7h_{ij}v_j) - v_k\partial_k(_7h_{ij}v_j) + \frac{1}{2}v_jv_k\partial_i(_7h_{jk})], \tag{19}$$

where $\bar{\rho}$ is the NS average density, $_7h_{ij} = \frac{32G}{45}x_k\varepsilon_{kl(i}J_{j)l}^{(4)}$ is the coefficient of the term of order c^{-7} in the expansion, and the even $_n\beta_j$ functions in Eq.(19) are shown to make it null contributions to the source dynamics (Rezzolla et al. 1999). The dependence on time-varying mass-quadrupole moments, of order $(v/c)^{5/2}$, is encoded in the term

$$\Delta F_i^{\frac{5}{2}\mathrm{PN}}(_7\alpha, _6\beta_i, _5h_{ij}) \equiv \bar{\rho}[-\partial_i(_7\alpha) - \partial_t(_6\beta_i) \tag{20}$$
$$+ v_j\partial_i(_6\beta_j) - v_j\partial_j(_6\beta_i) - \partial_t(_5h_{ij}v_j) - _5h_{ij}v_k\partial_kV_j],$$

with $_5h_{ij} = -\frac{4G}{5}\bar{I}_{ij}^{(3)}$ the coefficient of term of order c^{-5}.

From these equations, the equatorial component of the radiation reaction force can be recast in terms of the relevant current quadrupole as (LTV2001)

$$F_x^{\mathrm{GWs}} - iF_y^{\mathrm{GWs}} = -i|A|(x+iy)\left[3v_zJ_{22}^{(5)} + zJ_{22}^{(6)}\right], \tag{21}$$

while the axial contribution (along the Z-axis of spin) reads

$$F_z^{\text{GWs}} = -|A|\text{Im}\left[(x+iy)^2\left(3\frac{v_x+iv_y}{x+iy}J_{22}^{(5)}+J_{22}^{(6)}\right)\right]. \tag{22}$$

In the above equations the term $J_{22}^{(n)}$ corresponds to the nth time derivative of the time-dependent quadrupole current tensor J_{22}. In the non-linear evolution studied by LTV2001 the fluid velocity $\vec{v}$ was assumed to receive a small amplitude perturbation driven by the r-mode, being thus defined as

$$\vec{v} = \Omega_0\vec{\phi} + \alpha_0 R_0\Omega_0\left(\frac{r}{R_0}\right)^2\text{Re}(\vec{Y}_{22}^{B\star}), \tag{23}$$

where R_0 the radius of the non-spinning star, and $\Omega_0 = 0.635\sqrt{\pi G\bar{\rho}}$ is the angular frequency of the critical rotating model. In the regime of small initial r-mode amplitude, i. e., $\alpha_0 = 0.1$, with $\alpha \equiv \frac{8\pi R_0|J_{22}|}{\Omega_0\int\rho r^4 d^3x}$ (LTV2001), it was verified that the time-derivative of J_{22} is linear with the mode angular frequency ω. In fact, this result was shown to be accurate up to higher order time derivatives, that is: $J_{22}^{(n)} = (i\omega)^n J_{22}$, where the angular frequency in this non-linear regime reads: $\omega = -\frac{|J_{22}^{(1)}|}{J_{22}}$.

6.2. Funneling of the r-modes GWs Emission

If the back-reaction of GWs released is to kick the nascent pulsar, then the radiation must be beamed up to some level. As a large part of the rotation energy escapes as GWs from r-modes (LTV2001) the RRF axial (by definition parallel to the PNS spin) plus the equatorial components should act as the source of the funneling of the gravitational radiation emission, what is clearly seen from Eqs.(21-22). A rough estimate is obtained as the ratio of the RRF components Eqs.(21-22):

$$F_{x-iy} = -|A|\left[-3xyJ_{22}^{(5)}+3yv_xJ_{22}^{(5)}-2xyJ_{22}^{(6)}\right] \tag{24}$$

$$F_z = -|A|(-y)\left(3v_zJ_{22}^{(5)}+zJ_{22}^{(6)}\right). \tag{25}$$

To obtain order of magnitude estimates in all the computations henceforth we use $|J_{ij}| = M_{\text{NS}}R_{\text{NS}}^2|\vec{v}| \sim 10^{54}$ g cm^3 s^{-1} [$|\vec{v}|$ from Eq.(23)], with $M_{\text{NS}} = 1.44$ M$_\odot$, the NS mass, $R_{\text{NS}} = 17$ km, the star radius (LTV2001), and $\bar{\rho} = 1 \times 10^{14}$g cm^{-3}. The position $|x_i = (x,y,z)| = \bar{d} \sim R_{\text{NS}}$ defines the location at which the gravitational potential is being considered, and $\frac{d}{dt} = T_{dyn}^{-1} \equiv (R_{\text{NS}}^3/GM_{\text{NS}})^{1/2} = f_{\text{GWs}}^{-1} = P_0 \sim 1$ ms is the dynamical timescale, as inferred from the LTV2001 numerical simulations. For these parameters one can show that most of the GWs emission is funelled into a cone of aperture

$$\theta_{\text{GWs}}^{\text{beam}} = \tan^{-1}\left(\frac{F_z}{F_{x-iy}}\right) \sim \tan^{-1}\left(\frac{-4.8}{4.5}\right) \sim 47°, \tag{26}$$

which is enough to force the just born NS to move and rotate at birth since RRF is off-center applied. As it may happen that not all the pulsars can have their proper motion aligned with the spin axis (but see Weisskopf et al. 2000), then it is interesting to consider viable mechanisms to drive the r-mode off-centered. We discuss a bit this issue in the appendix.

7. Pulsar Kicks

7.1. Recoil Velocity

By using Eq.(15), the largest pulsar recoil velocity, $|\vec{V}_{rec}| \leq 1500$ kms^{-1}, as measured in surveys (Lyne & Lorimer 1994; Lorimer et al. 1993; Cordes & Chernoff 1998) can be obtained from the dynamical relation

$$
\begin{aligned}
L_{\mathrm{GW}} &= \frac{16G}{45c^7} \left| \left\langle J_{ij}^{(3)} J_{ij}^{(3)} \right\rangle \right|_{P=1\ \mathrm{ms}} \\
&\equiv F^{(\mathrm{back})}\, V_{rec} \bullet \cos\left(\vec{F}^{(\mathrm{back})}, \vec{V}_{rec} \right),
\end{aligned}
\tag{27}
$$

where $F^{(\mathrm{back})} = F^{7/2(PN)} \bar{R}_{\mathrm{NS}}^3$. This leads to

$$
V_{\mathrm{rec}} \sim 1500\,\mathrm{kms}^{-1} \left[\frac{\cos^{-1}(\vec{F}^{(\mathrm{back})}, \vec{V}_{rec})}{\pi\ \mathrm{rad.}} \right],
\tag{28}
$$

for values of the angle $\sim \pi$ rad between the GWs reaction force and the effective pulsar velocity. As can be seen from Eqs.(21-22), this relative angle is not fixed for all the cases, but rather vary from one pulsar to another because of its dependence on several factors as the RRF strength and the point of application, the fluid-velocity components, and the current quadrupole magnitude.

7.2. Natal Pulsar Periods and GWs Damping Timescale

GWs from r-modes spin down the pulsar rotation by carrying off angular momentum, J_i, at the total rate

$$
\begin{aligned}
\frac{dJ_i}{dt} &= -\left[\frac{5m\pi c^3}{2G} \right] f_{\mathrm{GW}} D^2 |h(t)|^2 \vec{e}_i \\
&= \left| \left(\frac{32G}{45c^7} \right) \left\langle J_{jl}^{(2)} J_{kl}^{(3)} \right\rangle \right| = \sum_A \varepsilon_{ijk} x_j^A F_k^{A(\mathrm{back})}.
\end{aligned}
\tag{29}
$$

To limit the NS's initial spin the RRF should be off-center applied at some place along its radius. At the centroid of the *force per unit length* distribution (the GW field on each quadrant in the plane wave approximation), it produces a torque which is proportional to that lever-arm ($x_j^A = \bar{d}$), i. e.:

$$
\vec{d} \times \vec{F}^{(\mathrm{back})} \equiv \bar{I}\alpha,
\tag{30}
$$

with $\bar{I} \sim M_{\mathrm{NS}} R_{\mathrm{NS}}^2$, and the induced angular acceleration

$$|\vec{\alpha}| \equiv \Delta\omega/\Delta t \sim \omega/T_{dyn}. \tag{31}$$

Assuming $\bar{d} = R_{NS} \sim 17$ km (LTV2001), nearly a tenth of a full GW wavelength, or the GW phase: $\varphi_{GW} = 2\pi/10 \longrightarrow \lambda_{GW}^{r-mode}/10$, then the *natal* pulsar spin frequency:

$$\bar{\omega} \sim 5.3 \times 10^3 \text{rad s}^{-1} \longrightarrow f_{GW} \sim 980\text{Hz} \longrightarrow P_0 \leq 1\text{ms}, \tag{32}$$

as computed by LTV2001. This result also agrees with the constraint posed by Cordes & Chernoff (1998). The lower initial spins already observed may be obtained for a lever-arm $\bar{d}$ half the one used here.

Finally, this RRF must drive the NS dynamics during the GWs damping timescale (Chau 1967):

$$\Delta T_{GW}^{(damping)} \sim 0.40 \text{ s} \left(\frac{1.4M_\odot}{M_{NS}}\right) \left(\frac{10^6 \text{ cm}}{R_{NS}}\right)^2 \left(\frac{P_0}{1 \text{ ms}}\right)^4. \tag{33}$$

This time interval is in agreement with the estimated duration $\Delta T_{GW} \sim 0.32$ s of the four thrusts expected to be given to the PNS in the very general kick mechanism discussed by Spruit & Phinney (1998).

As a summary, the above arguments allows one to conclude that the GWs radiation reaction kick mechanism provides a consistent picture that exhibits a clear concordance with observed rotation periods and drift velocities of runaway pulsars, and as such it constitutes a realization of the Nazin & Postnov (1997) conjecture.

8. Conclusions

In a series of papers Kusenko & Segrè (1999, a full review) proposed that the pulsar kick mechanism could be associated to the momentum asymmetry induced by neutrino oscillations. However, for neutrinos to do the job an extremely high magnetic field B$\sim 10^{15}$ G should be endowed by the young PNS. A field such as this is several orders of magnitude larger than the one inferred from the surveys of running away pulsars: B$\sim 10^{12}$ G (Lorimer et al. 1993; Lyne & Lorimer 1994; Cordes & Chernoff 1998; DRR99), and of course larger than the critical QED limit: $B_{crit} \sim 4 \times 10^{13}$G. Thus if one stands on the pulsar surveys it becomes clear that the linear momentum of the escaping neutrinos during the SN core collapse cannot be responsible for the pulsar kicks. I have shown this to be the case in a different realization of this GWs-Kick mechanism (Mosquera Cuesta 2002).

From the theoretical point of view, there is strong evidence for acceleration of black holes that have been impinged by a powerful GW (Brandt & Anninos 1999) (the reader should be awared that theoretical predictions of this rocket-like thrust go back to Bekenstein 1973). If such strong GWs are able to kick a BH to a velocity $v_{BH} \sim [150 - 250]$ km s^{-1}, as discussed in Brandt & Anninos (1999), then it is foreseeable that a similar acceleration impulse can be imparted to the nascent NS during its very early evolution after the definitive contraction to its characteristic size that follows the fading of the neutrino burst), leading to GWs kicks at birth. If the proposed mechanism actually operates when the pulsar is born, then it would turn the runaway pulsars the most compelling evidence for the existence of

GWs. Besides, the overwhelming population of high velocity pulsars would make it of the Einstein's theory of gravitation the most likely realized in nature.

9. Appendix: Mechanisms of Symmetry Breaking in Neutron Stars

How to turn the r-mode rotation axis off-centered in such a way that no azimuthal averaging can occur after thousands of NS turns? In other to provide a consistent justification for the occurrence in a natural way of this symmetry breaking, we focus next on a well-known phenomenon concerning spinning NSs which is in principle able to drive NS r-modes off-centered. It is likely that other effects can play some rôle in driving the r-mode axis off-centered, also.

9.1. Secular and Dynamical Instability of Rotating NSs in Newtonian Gravity

Firstly, we recall that young, hot, rapidly rotating NSs can break their symmetry if the ratio of kinetic to potential energy (T/W) exceeds a critical value, and the (l = 2, m = 2) bar-mode becomes unstable in a timescale: $\Delta T_{b-\text{mode}}^{\text{growth}}$ of nearly ten seconds. Whenever the viscosity of its constituent matter is low the star may be sensitive to the CFS instability, which is driven by gravitational radiation reaction, as quoted above. As concerns to rigidly rotating, homogeneous and self-gravitating fluid bodies, a Newtonian description of such stars allows to show that for a moderately spin frequency the star equilibrium configuration becomes an axisymmetric Maclaurin spheroid. For ratios $T/W \geq 0.2738$, the Maclaurin spheroid gets dynamically unstable. However, well before this limit be reached the so-called *secular* instability appears, where a couple of families of triaxial configurations set in for $T/W \geq 0.1375$. In the case of viscosity-driven instabilities, *kinetic energy is dissipated whereas angular momentum is conserved*. As a result, the fluid evolves along a sequence of equilibrium configurations close to a Riemann S ellipsoid towards to a *Jacobi* shaped one, which is the fluid arrangement having the lowest spin energy for a given angular momentum and mass of the intial Maclaurin configuration. Transitions to a *Jacobi* ellipsoid take place on the viscous timescale, which is longer than the dynamical one P_0.

9.2. General Relativistic Spontaneous Symmetry Breaking

By implementing a general relativistic description of the star fluid, Bonazzola, Frieben & Gourgoulhon (1998, BFG98) have generalized a former investigation on the effects of the space-time metric tensor scalar part to consider the influence of the so-called 3D-shift vector. Especifically, they studied the effect of general relativity in the viscosity-driven secular bar-mode instability of rapidly rotating NSs. The idea is to perturb an axisymmetric, stationary configuration and to study its evolution by constructing equilibrium sequences of triaxial quasi-equilibrium configurations, upon finding the solution for the dominant non-axisymmetric terms in the field equations (BFG98). Although their general treatment showed that the 3D-shift vector part turns out to inhibit the symmetry breaking in an efficient manner, it was found that the bar-mode instability can still develop for an astrophysically relevant mass of $M_{\text{NS}} \sim 1.4\,M_\odot$ when a stiff polytropic equation of state with adiabatic index $\gamma = 2.5$ is employed. The resulting triaxial NSs were shown to be potential targets for the GWs interferometric observatories LIGO, VIRGO, GEO-600 and TAMA-300.

The above arguments suggest that although the r- and bar-modes are of very different nature, they can affect significantly the evolution of each other so as to force, through the secularly viscosity-driven triaxiality, the r-mode to become off-centered all the way the bar-mode timescale. This way, the natural symmetry of the GWs-RRF along the PNS pointing axis is broken. This generates an effective linear momentum asymmetry enough to kick the pulsar. Because of the large GWs luminosity released we can see that a very tiny momentum asymmetry will suffice to impart large recoil velocities as those observed. We supplement this argument by quoting that according to Stergioulas and Font (2001), the possibility of a non-linear coupling of some stable fluid modes of NS pulsation (i.e., g-modes) and r-modes, in a growth time shorter than the GWs-RRF scale, has not been excluded yet.

Acknowledgments

I thank ICRA/CBPF for the PCI grant from the CNPq Institutional Capacitation Programme, and the Fundação de Amparo à Pesquisa do Estado do Rio de Janeiro (FAPERJ), Brazil, for the grant E-26/165.684/2003.

References

[1] Andersson, N., 1998, *Astrophys. J.* **502**, 708

[2] Friedman, J. L. and Morsink, S., 1998, *Astrophys. J.* **502**, 708

[3] Stergioulas, N. and Font, J., 2001, *Phys. Rev. Lett.* **86**, 1148

[4] Bildsten, L. & Ushomirsky, G., 2000, *Astrophys.* **529**, L33

[5] Blanchet, L., 1997, Phys. Rev. D 55, 714. See also Blanchet, L., Damour, T. & Schäfer, G., 1990, *MNRAS* **242**, 289

[6] Brandt, S. & Anninos, P., 1999, *Phys. Rev. D* **60**, 084005

[7] Bonazzola, S., Frieben, J. and Gourgoulhon, E., *Astron. Astrophys.* **331**, 280 (1998)

[8] Burrows, A., et al., 1995, *Astrophys. J.* **450**, 830

[9] Burrows, A. & Hayes, J., 1996, *Phys. Rev. Lett.* **76**, 352

[10] Chau, W.Y., 1967, *Astrophys. J.* **147**, 667

[11] Cordes, J. M. & Chernoff, D. F., 1998, *Astrophys. J.* **505**, 315

[12] Cowsik, R., 1998, *Astron. Astrophys.* **340**, L65

[13] Deshpande, A.A., Ramachandran, R. & Radakrishnan, V., 1999, *Astron. & Astrophys.* **351**, 195

[14] Kusenko, A. & Segrè, G., *Phys. Rev. Lett.* **77**, 4872 (1996), *Phys. Lett.* **B396**, 197 (1997).

[15] Akhmedov, E., Sciama, D. W. & Lanza, A., Resonant neutrino spin-flavor precession and pulsars kicks, *Phys. Rev.* **D56**, 6117 (1997).

[16] Grasso, D., et al., *Phys. Rev. Lett.* **81**, 2412 (1998).

[17] Nardi, E. & Zuluaga, J. I., Pulsar acceleration by asymmetric emission of sterile neutrinos, astro-ph/0006285, 20 June (2000).

[18] Harrison, E.R. & Tademaru, E., Acceleration of pulsars by asymmetric radiation, *Astrophys. J.* **201**, 447 (1975).

[19] Lai D., Chernoff D. & Cordes, J. M., 2000, *Astrophys. J.* **549**, 1111

[20] Mosquera Cuesta, H. J.,2000, *Astrophys. J. Lett.* **544**, L61

[21] Mosquera Cuesta, H. J., *Phys. Rev. D* **65**, 061503(R) (2002)

[22] Mosquera Cuesta, H. J. & K. Fiuza, *Eur. Phys. J.* **C 35**, 543-554 (2004)

[23] Lee Carson Loveridge, *Phys. Rev. D* **69**, 024008 (2004)

[24] A. L. Watts, N. Andersson, *MNRAS* **333**, 943 (2002) and references therein.

[25] N. Andersson, *Gravitational waves: A challenge for theoretical astrophysics*, Trieste, Italy, July 4-9, 2000. *ICTP Lecture Notes* Vol. 3, 297-314. See also N. Andersson, K. Kokkotas, *r-modes of Relativistic Stars,*in Living Reviews in General Relativity (2000).

[26] Janka H.-Th. & Müller E., 1994, *Astron. & Astrophys.* **290**, 290. See also Janka H.-Th. & Müller E., 1996, *Astron. & Astrophys.* **306**, 167

[27] Kaspi, V., et al., 1996, *Nature* **381**, 584

[28] A. Kusenko & G. Segrè, G., 1996, *Phys. Rev. Lett.* **77**, 4872. See alsoA. Kusenko & G. Segrè, 1999, *Phys. Rev.* **D59**, 061302

[29] L. Lindblom, Gravitational waves: A challenge for theoretical astrophysics, Trieste, Italy, July 4-9, 2000. *ICTP Lecture Notes* Vol. 3, 257-275

[30] Lindblom, L., Owen, B. & Morsink, S., 1998, *Phys. Rev. Lett.* **80**, 4843

[31] Lindblom, L., Tohline, J. & Vallisneri, M., 2001, *Phys. Rev. Lett.* **86**, 1152

[32] Lorimer, D.R., et al., 1993, *MNRAS* **263**, 403

[33] Lyne, A.G. & Lorimer, D.R., 1994, *Nature* **369**, 127

[34] Müller, E. & Janka, H.-Th., 1997, *Astron. &. Astrophys.* **317**, 140

[35] Camilo, F.., et al., 1996, *Astrophys. J.* **469**, 819

[36] Owen, B. & Lindblom, L., report astro-ph/0111024

[37] Nazin, S. & Postnov, M., 1997, *Astron. & Astrophys.* **317**, L79

[38] Nice D. J., et al., 1995, *Astrophys. J.* **449**, 156

[39] Rezzolla L., Lamb F. K. & Shapiro, S. L., 2000, *Astrophys. J.* **531**, L139

[40] Rezzolla, L., et al., 1999, *Astrophys. J.* **525**, 935

[41] Spruit, H. & Phinney, S.E., 1998, *Nature* **393**, 139

[42] Taylor, J. H., 1993, *AAS* **184**.680

[43] Weisberg, J. M. Taylor, J. H., 2003. *The Relativistic Binary Pulsar B1913+16*. Conference Proceedings, Vol. 302. Held 26-29 August 2002 at Mediterranean Agronomic Institute of Chania, Crete, Greece. Edited by Matthew Bailes, David J. Nice and Stephen E. Thorsett. San Francisco: Astronomical Society of the Pacific, 2003. ISBN: 1-58381-151-6, p.93

[44] Yoshida, Shin'I., et al., 2000, *MNRAS* **316**, 917

[45] Shijun Yoshida (Waseda U.), Shin'ichirou Yoshida (Wisconsin U., Milwaukee), Yoshiharu Eriguchi (Tokyo U., Astron. Dept.). "r-mode oscillations of rapidly rotating barotropic stars in general relativity: analysis by the relativistic cowling approximation ", Jun 2004. 9pp. *Mon.Not.Roy.Astron.Soc.* in press (2004)

[46] M. C. Weisskopf, et al. *Astrophys. J.* **536**, L81 (2000), Discovery of Spatial and Spectral Structure in the X-Ray Emission from the Crab Nebula

In: Pulsars: Discoveries, Functions and Formation
Editor: Peter A. Travelle, pp. 17-33
ISBN 978-1-61122-982-0
© 2011 Nova Science Publishers, Inc.

Chapter 2

CHANGES OF THE ORBITAL PERIODS OF THE BINARY PULSARS

Kostadin Trenčevski
Institute of Mathematics, Faculty of Natural Sciences
and Mathematics, Ss. Cyril and Methodius University,
P.O.Box 162, 1000 Skopje, Macedonia

Abstract

In this chapter the phenomena which yield to changes of the orbital periods of any binary systems are considered, and compared with the observed values for some binary pulsars. In the introduction the known formula caused by the gravitational radiation according to the general relativity, and the corrections caused by kinematic influence are presented. The method of evaluation of the decay of the orbital period via the advance of the periastron rate and the time dilation/gravitational red shift amplitude γ, is also presented. In the sections 2 and 3 are presented some recent results about time dependent gravitational potential, which also cause change of the orbital period on any binaries. It is assumed that the gravitational potential in the universe changes linearly, or almost linearly, with the time via $\Delta V = -c^2 H \Delta t$. This change is determined such that it is compatible with the Hubble red shift. It causes some deformations of the trajectories of the planetary orbits as well as the orbits of any binaries, such that the trajectories are not axially symmetric. The basic influence is the increasing the orbital period of any binary system, such that $\dot{P}_b = P_b H / 3$. In section 4 are considered the changes of the orbital periods of four pulsars according to the gravitational radiation together with the influence from the universe, and it is compared with the observed change of the orbital periods. In case of the Hulse-Taylor pulsar B1913+16 the dominant role has the gravitational radiation and it confirms very well; in case of the pulsar B1885+09 the dominant role has the influence from the gravitational potential and it confirms, and in case of the pulsar B1534+12 simultaneously confirm both the gravitational radiation and the gravitational potential in the universe. In the section 5 are considered some additional arguments about the changes and observations of the distances, times and velocities, caused via the gravitational potential in the universe. Using these arguments, in section 6 is discussed the correlation between the change of the distance Earth-Moon, the increasing of the lunar orbital period and the increasing of the average length of day.

1. Introduction

In 1993 the Nobel Prize in Physics was awarded to Russell Hulse and Joseph Taylor of Princeton University, for their discovery in 1974 of the binary pulsar known as PSR 1913+16. Later many new such binaries were discovered and a lot data were obtained. Among the main contributions to the science are the new tests of the general relativity that were done. Then appeared the first tests of the general relativity away from the solar system.

In this introductory section we present briefly the basic knowledge about these tests. The tests due mainly to the gravitational radiation, and later the geodetic precession was considered [11, 29]. We shall consider the tests about the gravitational radiation. The general relativity predicts a lost of energy under the gravitational radiation, which yields to decreasing of the relative semi-major axis of the orbit and as well as the orbital period P_b of the binary system, in our case a binary pulsar.

Five Keplerian parameters: projected semi-major axis $a_p \sin i$, eccentricity e, reference epoch T_0, binary orbital period P_b and argument of periastron w_0, as well as the three relativistic parameters means advance of the periastron rate $\langle \dot{w} \rangle$, time dilation/gravitational red shift amplitude γ and orbital change rate data $\dot{P}_b$ are directly measured. The other parameters are deduced from the previous parameters, assuming that the general relativity is true. Specially the masses of the pulsar and its companion, m_p and m_c, can be deduced from the previous parameters. Indeed, both masses can be calculated via the known relativistic formulae for the advance of the periastron shift and the dilation/gravitational red shift, given by

$$\langle \dot{w} \rangle = 3G^{2/3}c^{-2}(P_b/2\pi)^{-5/3}(1-e^2)^{-1}(m_p+m_c)^{2/3}, \tag{1}$$

and

$$\gamma = G^{2/3}c^{-2}(P_b/2\pi)^{1/3}em_c(m_p+2m_c)(m_p+m_c)^{-4/3}, \tag{2}$$

as a system of two equations with two unknowns. Further can be calculated the other orbital parameters like $\sin i$ and the orbital semi-major axes of the pulsar, its companion and the relative semi-major axis. Now we are interesting only for both masses m_p and m_c in order to obtain the relativistic prediction about the decreasing of the orbital period, and than to compare with the observed value. In case of a pair of collapsed stars, i.e. stars with small radiuses, the formula about the decreasing of the orbital period referes to the results of P.C.Peters and J.Mathews [15] and it is given by

$$\dot{P}_b = -\frac{192\pi G^{5/3}}{5c^5}\left(\frac{2\pi}{P_b}\right)^{5/3}(1-e^2)^{-7/2}\left(1+\frac{73}{24}e^2+\frac{37}{96}e^4\right)\frac{m_p m_c}{(m_p+m_c)^{1/3}}. \tag{3}$$

Notice that besides the post-Keplerian parameters $\langle \dot{w} \rangle$, γ, and $\dot{P}_b$, also the post-Keplerian parameters s for the shape of Shapiro delay, and r for the range of Shapiro delay are directly measured, such that we have overdetermined system of parameters. The observations show that these parameters r and s fit well with the other parameters, specially with $\langle \dot{w} \rangle$ and γ, and they are indeed confirmation tests of the general relativity.

Besides the previous change (3) of the pulsars orbital period, there appears also an effect of kinematic nature, which causes to observe a modified change of the orbital period. Indeed this is a result of the apparent acceleration due to the proper motion of the binary system. This effect is studied in [5], and it is given by

$$\left(\frac{\dot{P}_b}{P_b}\right)_{gal} = -\frac{a_z \sin b}{c} - \left(\frac{v_0^2}{cR_0}\cos l + \frac{v_1^2}{cR_1}\cos\lambda\right) + \mu^2\frac{d}{c}, \qquad (4)$$

where l and b are the galactic coordinates of the pulsar, a_z is the vertical component of the galactic acceleration, v_0 circular velocity at the Sun's galactocentric radius R_0, v_1 and R_1 are the corresponding values at the pulsar's location, μ proper motion, λ the angle between the Sun and the galactic center as seen from the pulsar, and d is the distance from the pulsar to the Sun. Some of the data in our galaxy are not very precise, and the main problem comes in determination of the distance d. The distance d can be estimated from the dispersion measure [21].

The comparison between the observed and predicted values of $\dot{P}_b$ via (3) and (4) will be done in section 4. First in sections 2 and 3 will be studied an additional change of the orbital period, which comes from the universe, and which is closely related with the Hubble constant H.

2. Change of the Gravitational Potential in the Universe and the Hubble Red Shift

In the recent paper [24] was developed an idea about linear (or almost linear) change of the gravitational potential in the universe. In this section, and also in sections 3 and 4 we present the results obtained in [24].

The assumed time dependent gravitational potential, which shall be considered in this section, is apart from the change of the gravitational potential near the massive bodies. We assume that the sign of the gravitational potential V is chosen such that the gravitational potential V is larger near the massive bodies. Under this choice of the sign of the gravitational potential V, if a light signal starts from a star with a frequency v_0, then according to the general relativity after time t its frequency will be

$$v = v_0\left(1 + \frac{t}{c^2}\frac{\partial V}{\partial t}\right).$$

Specially, on a distance $R = ct$ its frequency will be

$$v = v_0\left(1 + \frac{R}{c^3}\frac{\partial V}{\partial t}\right).$$

On the other side, according to the Hubble law, this frequency is

$$v = v_0\left(1 - \frac{RH}{c}\right),$$

where H is the Hubble constant, $H \approx 70$ km/s/Mpc. According to the last two equalities, we obtain that

$$\frac{\partial V}{\partial t} = -c^2 H \approx -2 \times 10^3 \frac{\text{cm}^2}{\text{s}^3}. \qquad (5)$$

Hence accepting this linear change, which is given by (5), the first and the most simple application appears the explanation of the Hubble red shift.

Since $\frac{\partial V}{\partial t} < 0$, in the past the time in the universe was slower than now, and in future it will be faster than now. Indeed, each second the time is faster approximately $k = (1 + H \times 1s) \approx 1 + \frac{1}{4.42 \times 10^{17}}$ times, assuming that $H \approx 70$ km/s/Mpc.

The reason of this change of the gravitational potential is probably in the change of the density of the dark energy, which is about 67% in the universe [7], but the following questions are open: Does the change is linear or almost linear, for example exponential? Does the potential changes everywhere in the universe by the same law (5), or it has a small dependence from the position in the universe?

The change of the gravitational potential on short time intervals happens very often. For comparison, the change of the Earth's gravitational potential in a lift which moves with velocity 2 cm/s away from the ground is approximately equal to $-c^2 H$, i.e. corresponds to the change of the gravitational potential in the universe. This change of the potential in the universe is about ten times smaller than the average change of the gravitational potential of the Earth on its orbit from the perihelion to the aphelion in a period of six months. The motion of the solar system in our galaxy also yields to almost linear change of the gravitational potential. This change appears as a constant for long time, but its intensity is much smaller than that which happens globally in the universe.

Note that the influence of the Doppler effect to the red shift is much smaller than the influence from the time dependent gravitational potential. Thus the red shift is only approximately proportional to the distance of the considered galaxy and now the velocities among the galaxies are not too large.

The presented hypothesis would be meaningless if it does not have some useful consequences. Besides the Hubble red shift, in the next sections we shall consider some other consequences. Moreover, it is very important that the linear change of the gravitational potential does not contradict with the general relativity, specially with the classical experiments (displacement of the spectral lines, Shapiro time delay, deflection of the light rays near the Sun, perihelion precession of the planetary orbits [30]), which confirm the general relativity. It shall be considered later.

One of the most important consequence of the considered hypothesis is the deformations of the planetary orbits, the orbits of the binary stars and the change of the orbital periods. It will be considered in the next section.

3. Deformation of the Orbits and Increasing of the Orbital Periods of any Binaries

Let us denote by X, Y, Z, T our natural coordinate system, in the deformed space-time, and let us denote by x, y, z, t the normed coordinates of an imagine coordinate system, where the space-time is "uniform", except close to the gravitational objects. This can be achieved by dividing each small space and time interval by a corresponding coefficient according to the general relativity and to see its length in case of vanishing of the gravitational potential. In this way we can not obtain a global coordinate system. Indeed, the coordinate system x, y, z, t is not adoptable, because x, y, z, t may not always be functions of X, Y, Z, T, and hence it should be conceived just like an orthonormal frame, i.e. a system of 4 orthonormal vectors. But however, we make a convention to call x, y, z, t coordinates also. We choose an arbitrary moment $t = T = 0$ as an initial moment for measuring the time. According to the general relativity we have the following equalities

$$dx = \left(1 + \frac{V}{c^2}\right)^{-1} dX = \left(1 + tH\right) dX, \qquad (6)$$

$$dy = \left(1 + \frac{V}{c^2}\right)^{-1} dY = \left(1 + tH\right) dY, \qquad (7)$$

$$dz = \left(1 + \frac{V}{c^2}\right)^{-1} dZ = \left(1 + tH\right) dZ, \qquad (8)$$

$$dt = \left(1 - \frac{V}{c^2}\right)^{-1} dT = \left(1 - tH\right) dT. \qquad (9)$$

Now the Hubble red shift is explained directly from (9). It is sufficient to assume that (6), (7), (8), and (9) are satisfied, and then it is not necessary to speak about the time dependent gravitational potential. Indeed, we accept (6), (7), (8), and (9) as axioms. The coefficients $1 + tH$ and $1 - tH$ probably should be exponential functions, but neglecting H^2 and smaller quantities we accept these linear functions. Thus the coefficient $1 + tH$ can be replaced by $1 + TH$.

From (6), (7), (8), and (9) we obtain

$$\left(\frac{dX}{dT}, \frac{dY}{dT}, \frac{dZ}{dT}\right) = \left(\frac{dx}{dt}, \frac{dy}{dt}, \frac{dz}{dt}\right)(1 - 2tH) \qquad (10)$$

and by differentiating this equality by T we have

$$\left(\frac{d^2X}{dT^2}, \frac{d^2Y}{dT^2}, \frac{d^2Z}{dT^2}\right) =$$

$$= \left(\frac{d^2x}{dt^2}, \frac{d^2y}{dt^2}, \frac{d^2z}{dt^2}\right) - 3tH\left(\frac{d^2x}{dt^2}, \frac{d^2y}{dt^2}, \frac{d^2z}{dt^2}\right) - 2H\left(\frac{dX}{dT}, \frac{dY}{dT}, \frac{dZ}{dT}\right). \qquad (11)$$

In normed coordinates x, y, z, t there is no acceleration caused by the time dependent gravitational potential, i.e. we can put there $H = 0$. Thus, in our natural coordinates X, Y, Z, T appears an additional acceleration

$$-3tH\left(\frac{d^2x}{dt^2}, \frac{d^2y}{dt^2}, \frac{d^2z}{dt^2}\right) - 2H\left(\frac{dX}{dT}, \frac{dY}{dT}, \frac{dZ}{dT}\right).$$

This acceleration appears also in the planetary and pulsars orbits. To study it more carefully, first we shall consider two special cases of deformations, as basic generators of deformations. In order to emphasize the new phenomena, we shall neglect the relativistic deformations, for example relativistic periastron precession. It means that in the normed coordinates we assume that the trajectories are Keplerian. Three parameters of the planetary orbits will be considered: 1. the angle between the radii given by the periastron and apastron, or its departure $\Delta\varphi$ from π; 2. the periastron precession; and 3. the quotient $\Theta_2 : \Theta_1$, where Θ_1 and Θ_2 are two consecutive orbital periods.

(i) Since x, y, z, t are not functions of X, Y, Z, T, in order to see the influence of the constant H, we are looking for an adopted transformation

$$x = (1 + \lambda tH)X, \ y = (1 + \lambda tH)Y, \ z = (1 + \lambda tH)Z, \ t = T, \ (\lambda = const.), \qquad (12)$$

which means that we have only change in the space coordinates, but there is no change in the time coordinate. Using the normed coordinates x, y, z, t the trajectory in the plane of motion is given by

$$\frac{1}{r} = \frac{\rho_1 + \rho_2}{2} + \frac{\rho_1 - \rho_2}{2} \cos \varphi, \qquad (13)$$

where $\rho_1 = 1/r_1$ and $\rho_2 = 1/r_2$ are constants, known from the Newtonian theory.

According to (12), the equality (13) becomes

$$\frac{1}{R} = (1 + \lambda T H) \left[\frac{\rho_1 + \rho_2}{2} + \frac{\rho_1 - \rho_2}{2} \cos \varphi \right]. \qquad (14)$$

The angle φ is a common parameter for both coordinate systems. The equation $\frac{d\rho}{d\varphi} = 0$ for the extreme values of $1/R$, according to (14), yields to

$$\frac{d}{d\varphi} \left[\frac{\rho_1 + \rho_2}{2} (1 + \lambda t H) + \frac{\rho_1 - \rho_2}{2} (1 + \lambda t H) \cos \varphi \right] = 0,$$

$$-\frac{\rho_1 - \rho_2}{2} (1 + \lambda t H) \sin \varphi + \left[\frac{\rho_1 + \rho_2}{2} + \frac{\rho_1 - \rho_2}{2} \cos \varphi \right] \frac{dt}{d\varphi} \lambda H = 0,$$

$$\sin \varphi = \left[\frac{\rho_1 + \rho_2}{\rho_1 - \rho_2} + \cos \varphi \right] \frac{\lambda H}{\frac{d\varphi}{dt}}.$$

The solution of this equation of φ when $\varphi \approx 0$ can be approximated by putting $\cos \varphi = 1$ and hence

$$\varphi_1 = \frac{2\rho_1}{\rho_1 - \rho_2} \frac{\lambda H}{(\frac{d\varphi}{dt})_{per.}} = \frac{2r_1 \lambda H}{(\rho_1 - \rho_2)C}, \qquad (15)$$

where $C = r^2 \frac{d\varphi}{dt} = const.$ according to the second Kepler's law. The solution of φ when $\varphi \approx \pi$ can be approximated by putting $\cos \varphi = -1$, i.e.

$$\varphi_2 = \pi - \frac{2\rho_2}{\rho_1 - \rho_2} \frac{\lambda H}{(\frac{d\varphi}{dt})_{aph.}} = \pi - \frac{2r_2 \lambda H}{(\rho_1 - \rho_2)C}. \qquad (16)$$

Hence we obtain

$$\varphi_2 - \varphi_1 = \pi - \frac{2(r_2 + r_1)\lambda H}{\frac{r_2 - r_1}{r_1 r_2}C} = \pi - \frac{2r_1 r_2 \lambda H}{Ce},$$

where e is the eccentricity of the orbit. According to the Kepler's orbit it is $\frac{r_1 r_2}{C} = \frac{\Theta\sqrt{1 - e^2}}{2\pi}$, where Θ is the orbital period, and thus

$$\Delta \varphi = \varphi_2 - \varphi_1 - \pi = -\frac{\lambda H \Theta \sqrt{1 - e^2}}{\pi e}. \qquad (17)$$

Further, according to (15) for the precession of the periastron we obtain the angle

$$\frac{2\rho_1}{\rho_1 - \rho_2} \left[\frac{\lambda H}{(\frac{d\varphi}{dt})_\Theta} - \frac{\lambda H}{(\frac{d\varphi}{dt})_0} \right] = 0, \qquad (18)$$

neglecting the terms of order H^2. Hence the angle from the apastron to the periastron is equal to

$$\pi + \frac{2r_1 r_2 \lambda H}{Ce} = \pi + \frac{\lambda H \Theta \sqrt{1-e^2}}{\pi e},$$ (19)

and the orbit is not axially symmetric.

Since $t = T$ in this case, it is $\Theta_1 = \Theta_2$, i.e.

$$\Theta_2 : \Theta_1 = 1.$$ (20)

It is easy to check that the accelerations in both coordinate systems are related by

$$\left(\frac{d^2 X}{dT^2}, \frac{d^2 Y}{dT^2}, \frac{d^2 Z}{dT^2}\right) = (1 - \lambda TH)\left(\frac{d^2 x}{dt^2}, \frac{d^2 y}{dy^2}, \frac{d^2 z}{dt^2}\right) - 2\lambda H\left(\frac{dX}{dT}, \frac{dY}{dT}, \frac{dZ}{dT}\right).$$ (21)

(ii) Now assume that instead of (6), (7), (8), and (9), we have the following coordinate transformations

$$x = X, \quad y = Y, \quad z = Z, \quad dt = (1 - \mu tH)dT, \quad (\mu = const.),$$ (22)

which means that we have only change in the time coordinate. Thus, the orbit in the (R, φ)-plane is the same ellipse like the ellipse according to the x, y, z, t coordinate system.

Using that $T = t + \frac{\mu}{2}Ht^2$, which obtains by integration of (22), for the quotient $\Theta_2 : \Theta_1$, which corresponds to $t_2 = t_1$ for the orbital periods, we obtain

$$\Theta_2 : \Theta_1 = \left([2\Theta + \frac{\mu}{2}H(2\Theta)^2] - [\Theta + \frac{\mu}{2}H\Theta^2]\right) : \left(\Theta + \frac{\mu}{2}H\Theta^2\right) = 1 + \mu\Theta H,$$

$$\Theta_2 : \Theta_1 = 1 + \mu\Theta H.$$ (23)

We notice that the equations (22) imply that

$$\frac{dX}{dT} = \frac{dx}{(1 + \mu Ht)dt} = \frac{dx}{dt}(1 - \mu tH),$$

$$\frac{d^2 X}{dT^2} = \frac{d}{(1 + \mu Ht)dt}\left(\frac{dx}{dt}(1 - \mu tH)\right) = (1 - \mu tH)^2\frac{d^2 x}{dt^2} - \mu H\frac{dX}{dT},$$

and analogously is true for Y and Z coordinates, i.e.

$$\left(\frac{d^2 X}{dT^2}, \frac{d^2 Y}{dT^2}, \frac{d^2 Z}{dT^2}\right) = (1 - 2\mu TH)\left(\frac{d^2 x}{dt^2}, \frac{d^2 y}{dt^2}, \frac{d^2 z}{dt^2}\right) - \mu H\left(\frac{dX}{dT}, \frac{dY}{dT}, \frac{dZ}{dT}\right).$$ (24)

Combining both special cases (i) and (ii), we obtain that if the acceleration is given by

$$\left(\frac{d^2 X}{dT^2}, \frac{d^2 Y}{dT^2}, \frac{d^2 Z}{dT^2}\right) =$$

$$= (1 - (\lambda + 2\mu)TH)\left(\frac{d^2 x}{dt^2}, \frac{d^2 y}{dt^2}, \frac{d^2 z}{dt^2}\right) - (2\lambda + \mu)H\left(\frac{dX}{dT}, \frac{dY}{dT}, \frac{dZ}{dT}\right),$$ (25)

then the angle $\Delta\varphi$ is given by (17), there is no precession of the periastron and the quotient $\Theta_2 : \Theta_1$ is given by (23). This happens for the composition of the transformations (12) and (22), i.e.

$$x = (1 + \lambda t H)X, \ y = (1 + \lambda t H)Y, \ z = (1 + \lambda t H)Z, \ dt = (1 - \mu t H)dT. \tag{26}$$

Let us return to our initial problem about the space-time in the universe. We determine the constants λ and μ such that the accelerations (25) and (11) are identical. Thus, we obtain $\lambda = \frac{1}{3}$, $\mu = \frac{4}{3}$, and there is no periastron precession caused by the time dependent gravitational potential. Hence the angle from the periastron to the apastron is $\pi - \frac{\lambda H \Theta \sqrt{1-e^2}}{e\pi} = \pi - \frac{H\Theta\sqrt{1-e^2}}{3e\pi}$ according to the observer where there is no time dependent gravitational potential, while the angle from the apastron to the periastron is $\pi + \frac{\lambda H \Theta \sqrt{1-e^2}}{e\pi} = \pi + \frac{H\Theta\sqrt{1-e^2}}{3e\pi}$ according to the same observer. This angle increases if the orbital period increases and/or the eccentricity decreases. Below in the Table are given the non-relativistic predicted values of $\Delta\varphi$ measured in milliarcseconds for each planet, assuming that H^{-1} is 14×10^9 years.

While the relativistic perihelion shift decreases with the distance, $\Delta\varphi$ increases for the planets which are on larger distances from the Sun. This phenomena is difficult to detect, not only because of the perturbations from the other planets, but this effect appears periodically $-\Delta\varphi, \Delta\varphi, -\Delta\varphi, \Delta\varphi, \cdots$, and does not accumulate with the time like for the periastron precession. Thus it is almost impossible this effect to be detected for the trajectories of the binary pulsars, where the orbital periods are often very small.

From (23) it follows that the quotient $\Theta_2 : \Theta_1$ of two consecutive orbital periods is equal to $1 + \mu \Theta H = 1 + \frac{4}{3} \Theta H$. This shows that each next orbit has a prolonged period for a factor $1 + \frac{4}{3}\Theta H$. But our time Θ is also prolonged for a factor $1 + \Theta H$ according to (9). Thus we observe that each next orbit is prolonged for the factor

$$\Theta_2 : \Theta_1 = \frac{1 + \frac{4}{3}\Theta H}{1 + \Theta H} = 1 + \frac{1}{3}\Theta H. \tag{27}$$

Indeed, each short time interval on the orbit should be divided by $1 + \frac{4}{3}\Delta T H$ in order to obtain the uniform time Δt in normed coordinates, while each short time of our clock should be divided by $1 + \Delta T H$ in order to obtain the homogeneous time Δt. This discrepancy of $\frac{1}{3}\Delta T H$ is a product of the nonholonomy coordinate system, since the system (6), (7), (8), and (9) is Pfaffian. This effect of increasing of the orbital period for the Earth is about 0.751 ms per year, but this is masked from the perturbations from the neighborhood. In the Table below are given also the differences $\Theta_2 - \Theta_1$ in seconds for all planets in the solar system.

The change of the gravitational potential causes negligible changes in the results of the known experiments about general relativity. Indeed, the effects from the time dependent gravitational potential are negligible for short time intervals, as for the experiments from general relativity, except for the perihelion precession. On the other hand we saw that the time dependent gravitational potential does not cause a perihelion /periastron shift.

Planet	$\Delta\varphi\,(10^{-3})''$	$\Theta_2 - \Theta_1\,(s)$
Mercury	-0.002	0.000,043,6
Venus	-0.16	0.000,285
Earth	-0.094	0.000,751
Mars	-0.032	0.002,66
Jupiter	-0.38	0.105,73
Saturn	-0.82	0.651,985
Uranus	-2.9	5.304,086
Neptune	-28.8	20.398,2
Pluto	-1.6	46.664,6

4. Analysis of the Prediced and Observed Values of the Change of the Orbital Periods of some Bynary Pulsars

In case of binary pulsars we shall use the notation P_b istead of Θ. From (27) it just follows that

$$\dot{P}_b = \frac{1}{3}P_b H. \tag{28}$$

While the gravitational radiation decreases the orbital period, (28) increases. The discrepancy which appears without using the formula (28) sometimes was explained by variation of the gravitational constant G, proposed by Dirac many years ago, but now that hypothesis is not needed.

We shall consider in this section four pulsars: B1913+16, 2127+11C, B1534+12 and B1885+09. It is interesting that for relativistic pulsars the orbital time is very small and then the gravitational radiation large, while the influence from (28) is small because of the short orbital period. Thus in these cases the gravitational radiation has dominant role, while the formula (28) has a minor role, and thus in such cases we have confirmation of the general relativity. This is the case of the pulsars B1913+16, 2127+11C. On the other side there is a pulsar (B1885+09), which has long orbital period such that the gravitational radiation is negligible, and the increasing of the orbital period according (28) has dominant role. In this case we have confirmation of the hypothesis about the time dependent gravitational potential. Finally we also have a "mixed case" (B1534+12), where each value of (3), (4), and (28) is not negligible. There we have simultaneous confirmations of the formulae (3) and (28). It should be mentioned that the pulsars B1534+12 and B1885+09 have very stable timings, and hence they are very convenient for precise observations.

The Hulse-Taylor pulsar B1913+16 [22, 28, 29] after 30 years of sequential observations enables with numerous and precise data for relativistic tests. Its orbital period is $P_b = 0.32299744893$ days and the observed decay of the orbital period is

$$(\dot{P}_b)_{obs} = -2.4184 \times 10^{-12} \text{ s/s}.$$

The precise measurements of the periastron shift and the parameter γ enables to find precise values of both masses: $m_p = 1.4414 \pm 0.0002$ and $m_c = 1.3867 \pm 0.0002$ solar masses [29]. This enables to find precise value of the orbital decay caused by the gravitational radiation, which is

$$(\dot{P}_b)_{GR} = -(2.40242 \pm 0.00002) \times 10^{-12} \text{s/s}.$$

According to the present knowledge of the galactic values, the values of (4) yields to [29]

$$(\dot{P}_b)_{Gal} = -(0.0128 \pm 0.005) \times 10^{-12} \text{s/s}.$$

According to (28) the increasing of the orbital period is given by [29]

$$\dot{P}_b = 0.021 \times 10^{-12} \text{s/s}.$$

Hence we see that the predicted value from the general relativity is in excellent agreement with the observations. The value predicted by (28) is only 0.875% of the value $(\dot{P}_b)_{GR}$ and it is 64% larger than $(\dot{P}_b)_{Gal}$. Having in mind the uncertainties of $(\dot{P}_b)_{Gal}$, mainly caused by the poor knowledge of the distance, the proper motion of the pulsar and the distance from the Sun to the center of the galaxy, we can not draw any conclusion for the validity of (28).

A similar situation we have with the relativistic binary pulsar B2127+11C [16] in the globular cluster M15. Its orbital period is $P_b = 0.335282$ days, and its decay of the orbital period is observed to be $\dot{P}_b = -3.937 \times 10^{-12}$ s/s [4]. The predicted value according to (28) is $\dot{P}_b = 0.0218 \times 10^{-12}$ s/s, which is only 0.55% from the observed value. On the other hand, since the pulsar is in a cluster and there appear unknown gravitational forces, a test of general relativity prediction can not be better than 2% [16]. Hence we can not draw again any conclusion about the validity of (28).

In case of PSR B1534+12, $P_b = 0.4207372993$ days, the decay of P_b caused by the gravitational radiation is about -0.1924×10^{-12} s/s, while the increasing caused by (28) is about 0.027×10^{-12} s/s, and hence together we have decay about -0.1654×10^{-12} s/s. On the other side, the observed value of $\dot{P}_b$ in [20] is $(-0.174 \pm 0.011) \times 10^{-12}$ s/s, where the galactic corrections (4) are included. This is in agreement of the measured and theoretical results. The measurements in [20] are done at the Arecibo Observatory, and some earlier measurements done with radio telescopes at Arecibo Observatory, Green Bank, and Jodrell Bank [19], show that the value of $\dot{P}_b$ is $(-0.167 \pm 0.018) \times 10^{-12}$ s/s, where the galactic corrections are included. Hence we have excellent agreement of the theoretical results via formulae (3), (4) and (28) from one side and the observed values from the other side.

The agreement is also very good for the pulsar B1885+09 [10, 17]. In [10] it is found that $P_b = 12.32717$ days and $\dot{P}_b = 0.6 \times 10^{-12}$ s/s. In this case the influence of the gravitational radiation is negligible, while the the kinematic galactic corrections are one order smaller than the observed value. Hence, $\dot{P}_b$ is caused mainly from (28). If this value $\dot{P}_b = 0.6 \times 10^{-12}$ s/s does not change in future, i.e. it is the true one, neglecting the kinematic galactic corrections, it is easy to obtain that $H = 52.4$ km/s/Mpc. Having in mind the previous uncertainties, this case also confirms the presented theory about the time dependent gravitational potential.

Unfortunately, we do not know another binaries where the formulae (3) and (28) can be tested. Indeed, for many binaries the observed value of $\dot{P}_b$ is not published, or not precisely measured yet, for example PSR J1518+4904 [14], PSR J1811-1736 [12], PSR B2303+46 [23], PSR B1820-11 [13, 23], PSR J0737-3039A [3], J1829+2456 [4], J1141-6545 [9], and PSR J0621+1002 [18]. Some of them PSR J1518+4904, PSR J1811-1736, PSR B2303+46,

PSR B1820-11, and PSR J0621+1002 having orbital periods respectively 8.634 days, 18.78 days, 12.34 days, 357.76 days, and 8.318 days, are very convenient for testing the formula (28), because of the long orbital periods.

5. Measurements of the Distances, Times, and Velocities

In this section we shall try to make further development of the presented theory about time dependent gravitational potential. The basic assumptions were sufficient to draw the consequences presented in sections 3 and 4, but however they are not sufficient to draw some conclusions about some essential questions tied with the measurements of the lengths and velocities. Such questions will be considered in this section, by adding new assumptions. In this section and section 6 they will be supported by another examples, like the Pioneer anomaly acceleration, increasing of the average length of day, increasing of the distance Earth-Moon, and increasing of the lunar orbital period.

We can not directly detect any shrinking/dilation in measuring the distances from the formulae (6), (7), (8), and (9), because our units of measurement also change with the time. If we measure the time as we saw at the end of section 3, the orbital period is longer for factor $1 + \frac{1}{3}\Theta H$. The basic reason for this effect comes from the fact that the "coordinate system" x, y, z, t is not adopted. Otherwise, this quotient will be 1. In this case it is important to underline that the time T is measured by the atomic clocks. It is confirmed with the examples of pulsars in section 4. If we want to use the time t, we can do that by norming each small time interval with the factor $1 - TH$.

Now let us consider the equality (10). It follows that each velocity decreases with the time by a factor $1 - 2tH$. It also happens with the velocity of light. This decreasing of the velocity of light corresponds to the Shapiro time delay effect from the general relativity. Thus we indirectly observe this effect via the Shapiro time delay. But from the basic equations (6), (7), (8), and (9), we can not conclude whether we are able directly to observe this change of the velocity of light. The following question appears. If we measure the velocity of light on short distances and short time intervals measuring with the same instruments, shall we measure the same value of c independently of time? Although the answer does not follow from (6), (7), (8), and (9), this practically can be checked. Indeed, the precise laser equipment measure the velocity of light better than 20 cm/s [31]. Thus, if the velocity of light is measured to decrease about 4 cm/s per year according to (10), then after 10 years this decreasing would be measurable. Since, such a decreasing with the precise laser equipment is not known, we accept that the assumption that the velocity of light is independent from the time. The constancy of the velocity of light is correlated by using atomic clock, and atomic clocks we accepted as clocks which measure the coordinate time T. Moreover, in October 1983 the velocity of light was declared as an absolute constant. Hence we can use the same notation c without missunderstanding. The previous discussion about the constancy of c is equivalent to the following experiment. Let us measure the velocity of light on the Earth when the Earth is at the perihelion, and later when the Earth is at the aphelion. In this case the change of the Sun gravitational potential is dominant, compared with the change of the gravitational potential in the universe. According to the assumed constancy of c, the velocity of the light will be measured unchanged using the same instruments.

Although we shall not measure any variation in the velocity of light, the change of the frequency can be detected. The formula for the Doppler effect, neglecting the relativistic corrections, in X, Y, Z, T coordinates is given by

$$\nu = \nu_0 \left[1 - H\frac{R}{c} - \frac{1}{c}\frac{dR}{dT} \right] =$$

$$= \nu_0 \left[1 - H\frac{R}{c} - \frac{1}{c}\frac{dr}{dt}\frac{1-tH}{1+tH} \right] = \nu_0 \left[1 - H\frac{R}{c} - \frac{1}{c}\frac{dr}{dt}(1-2tH) \right]. \tag{29}$$

The formula (29) will play basic role in the case of anomalous acceleration for spacecraft Pioneer 10 and 11 [1, 2]. The expression $H\frac{R}{c}$ gives the Hubble red shift mentioned in section 2. For motion with constant velocity $\frac{dr}{dt}$, we see that the frequency will decrease according to the factor $(1 - 2tH)$. In case of extremely large distances R, the expression $\frac{\nu}{c}(1 - 2tH)$ is negligible with respect to $H\frac{R}{c}$. This is a case with the red shift from the other galaxies, and we observe $\nu \approx \nu_0(1 - \frac{RH}{c})$.

Since many distances in the astrophysics are measured via the velocity of light in two directions, we shall discuss it now. Let us consider first a body with ends A and B. Although its length is decreasing with the time according to (6), (7), and (8), indeed it is a constant. Since c is a constant, the light will always pass the distance from A to B for the same time T. Moreover, if a radio signal comes form A to B, there will not appear Doppler effect, except that from the Hubble red shift, i.e. HR/c, such that the points A and B are in rest.

Now let us measure the distance from us to a body moving under the gravitation on an orbit. For the sake of simplicity we assume that it moves on a circle, but the result is more general. If the distance changes under the law $R = r(1 - tH)$, then there will not be observed change in the distance between the two points, as previously. But, since the distance changes under the law $R = r(1 - \lambda tH) = r(1 - \frac{1}{3}tH)$ according to section 3, it will be measured by laser that the distance between the two points is increased for the factor $1 + \frac{2}{3}tH$. The previous discussion also holds for the distances to the perihelion and aphelion of the orbit. The measured change in time will be according to the factor $1 + \frac{1}{3}tH$, as we saw in section 3.

The previous discussion yields that the data from the pulsars which we obtain now do not exactly correspond with the data from many years ago. If the pulsar is on a distance d, then this uncertainty is of order $\frac{dH}{c} \sim 10^{-6}$. This is negligible and can not cause any essential influence of the results.

The change of mass can not be detected directly, because parallel to the change of the mass we have change of the units of mass. Analogously as we assumed that the velocity of light, which connects the lengths and the times, is not detectable for measurements, we also assume that the gravitational constant G, which connects the lengths, times and masses is also undetectable for measurements. Thus we can use the constant G without any missunderstanding.

Now the previous discussion will be applied for a possible explanation of the anomalous acceleration a_P of spacecraft Pioneer 10 and 11. It was considered in [24], and now give better and improved explanation of the formula (34) in [24], i.e.

$$a_P = cH\left(\frac{7}{3}\frac{dR}{dT}\frac{T}{R} - 1\right). \tag{30}$$

Let $R = R(T)$ be the distance from the spacecraft to the DSN antenna and let us calculate $\frac{\Delta v}{v_0} = (v_{obs} - v_0)/v_0$. According to (29) we obtain

$$\frac{\Delta v}{v_0} = -2H\frac{R}{c} - 2\frac{\frac{dR}{dT}}{c} = -2H\frac{R}{c} - 2\frac{\frac{dr}{dt}}{c}(1 - 2Ht), \tag{31}$$

where we use factor 2 for the two way data. On the other side, since the authors of [1, 2] have not used the presented theory which includes the Hubble constant, they measure the anomalous acceleration a_P introduced by (15) from [1], i.e.

$$\left[v_{obs}(t) - v_{model}(t)\right]_{DSN} = -v_0\frac{2a_P R}{c^2}, \qquad v_{model} = v_0\left[1 - 2\frac{v_{model}(t)}{c}\right], \tag{32}$$

where v_0 is the reference frequency, v_{model} is the modeled velocity of the spacecraft due to the gravitational and other forces, and v_{obs} is the frequency of the re-transmitted signal observed by DSN antennae. Moreover, using the results in this section, and (32) we obtain

$$\frac{\Delta v}{v_0} = 2a_P\frac{R}{c^2} - 2\frac{\left(\frac{dR}{dT}\right)_{measured}}{c} = 2a_P\frac{R}{c^2} - \frac{2}{c}\frac{dr(1 + (1 - \lambda)tH)}{dt(1 + (\mu - 1)tH)} =$$

$$= 2a_P\frac{R}{c^2} - \frac{2}{c}\frac{dr}{dt}\left(1 + \frac{1}{3}tH\right). \tag{33}$$

From (31) and (33) we get

$$-2H\frac{R}{c} - 2\frac{\frac{dr}{dt}}{c}(1 - 2Ht) = 2a_P\frac{R}{c^2} - \frac{2}{c}\frac{dr}{dt}\left(1 + \frac{1}{3}tH\right),$$

and hence we obtain finally (30). This equality gives the required expression of the acceleration a_P as a function of the distance R. Now we are not able to determine the initial value of T, i.e. when we should start to measure the time T in (30). It depends on the initial conditions for the motion of the spacecraft, not necessary tied with its launch. Note also that for each initial value of T, when T tends to infinity, or R tends to infinity, then a_P tends to $\frac{4}{3}cH$. Thus, for different spacecraft are obtained close but different almost constant values of a_P, $a_P \approx \frac{4}{3}cH$. In an ideal case, if the experiment is done in an inertial system, then $dR/dT = R/T$ is a constant velocity and $a_P \equiv \frac{4}{3}cH$. Since $a_P \approx 8.74$ cm/s^2 [1, 2], for H we obtain $H = 67.465$ km/s/Mpc.

6. Increasing of the Distance to the Moon, Lunar Orbital Period, and the Length of Day

In the last few decades it was detected that the Moon goes away from the Earth about 3.8 cm per year [25] and it was detected that the orbital period of the Moon around the Earth is also increasing [6]. We shall use the last value of 0.019 s per century [26], which is known to the author. For both values 3.8 cm/year and 0.019 s/century we do not know the estimates of accuracy. Moreover it is measured that the length of the average day on the Earth increases about 1.7 ms per century [27]. This increasing is a consequence of the decreasing of the angular velocity of the Earth. The decreasing of the angular velocity is related with the increasing of the distance to the Moon and the increasing of the Moon's orbital period. One of the basic reason for these three phenomena is the tidal dissipation. In this section we shall show that applying the previous results, these three observations can be explained without large discrepancies. We agree to use the value of the Hubble constant $H = 1/14\text{Gyr} \approx 73$ km/s/Mpc.

First we give some draft explanations of the increasing of the average length of day (LOD) until now. Let us consider the simplified system of the Earth and Moon assuming that the Moon's orbit is a circle, and let R is the distance Earth-Moon, M_e and M_m are the masses of the Earth and Moon, $M = M_e M_m/(M_e + M_m)$, r is the radius of the Earth, and Ω is the Moon's orbital angular velocity. We use the conservation law of the angular moment

$$J = MR^2\Omega + \frac{2}{5}M_e r^2 w = M\sqrt{G(M_e + M_m)R} + \frac{2}{5}M_e r^2 w = const.,$$

where the angular momentum of the Moon is omitted since it is rather smaller than the other summands. By differentiating this equality we get

$$\frac{dR}{dt} = -\frac{4}{5}\frac{M_e r^2 \sqrt{R}}{M\sqrt{G(M_e + M_m)}}\frac{dw}{dt}. \tag{34}$$

Hence the decreasing of the Earth's angular velocity $\frac{dw}{dt}$ is proportional to the increasing of the Moon's distance R. Assuming that $dR/dt = 3.8$ cm/yr, we find that $dw/dt \approx -4.65 \times 10^{-22}$ rad/s^2. More precise calculations, using satellites measurements, predict that [8]

$$\left(\frac{dw}{dt}\right)_{tidal} \approx -(6.1 \pm 0.4) \times 10^{-22}\text{rad/s}.$$

This implies that the average increasing of the LOD is about 2.3 ms per century. Indeed, this is induced by the tidal dissipation. The discrepancy of 2.3-1.7=0.6 ms in LOD explains by the non-tidal acceleration, post-glacial rebound, which causes a net transfer of mantle material toward the poles. It decreases the Earth's polar moment of inertia and give rise to the non-tidal acceleration of the Earth.

Now we give the explanation according to the results in the previous sections. The distance Earth-Moon changes, and we measure an increasing of this distance for

$$\frac{2}{3}H \times (384.4 \times 10^8\text{cm}) \times 1\text{yr} = 1.83\text{cm},$$

i.e. 1.83 cm per year. The rest part of 3.8-1.83=1.97 cm dues to the tidal acceleration. The increasing of the distance of 1.83 cm/year caused by the change of the gravitational potential in the universe has no influence to the angular velocity of the Earth and it can be conceived only a result of our choice of the coordinate system, because it disappears in normed coordinates. Thus, only the increasing of 1.97 cm/year has influence on w. Assuming a linear dependence between the LOD and dR/dt, now we estimate that the length of average day increases for $\frac{2.3 \times 1.97}{3.8}$ ms = 1.19 ms per century, caused by the tidal dissipation. Further we shall determine the increasing of the length of average day caused by the time dependent gravitational potential. According to x,y,z,t coordinates it is $\frac{d\varphi}{dt} = const.$, which means that $\frac{d\varphi}{(1-tH)dT} = const.$, i.e. $\frac{d\varphi}{dT} = const.(1 - tH)$, because the angle φ

is a common parameter in both coordinate systems. This decreasing of the angular velocity implies increasing of the length of an average day, such that each next day is longer for factor $1 + TH$, where T is the length of one day. This implies that after one century the day will be longer

$$\frac{1}{14 \cdot 10^9 \text{y}} \times 86400\text{s} \times 100\text{y} = 0.617\text{ms}.$$

Both increasings, according to the tidal dissipation and the gravitational potential yield to 1.19+0.61= 1.8 ms per century. In this case we have only a slight decreasing about 0.1 ms per century caused by the nontidal accelerations. It is one order smaller than the increasing caused by the tidal dissipation.

Further, let us consider the value of increasing of the orbital period of the Moon. First we shall find the increasing of the orbital period caused by the tidal acceleration. By differentiating the equality $R^3 = T^2 \times const.$ we obtain $\frac{3}{2}\frac{\Delta R}{R} = \frac{\Delta T}{T}$, where R and T are the distance to the Moon and orbital period of the Moon, and ΔR and ΔT are the corresponding increasing per year. Using that $\Delta R = 1.97$ cm, $R = 384400$ km, and for the sidereal month $T \approx 2360000$ s, i.e. about 27.32 days, we obtain the increasing $\Delta T = 0.018$ s per century. This is very close to the measured value of 0.019 s. The influence from the gravitational potential is equal to 0.0056 s. For both together we obtain 0.018+0.0056=0.0236 s per century.

If we do not use the results from the previous sections, i.e. if we neglect the time dependent gravitational potential, then the increasing of 3.8 cm/year corresponds to $\Delta T = \frac{3}{2} \times 2360000\text{s} \times \frac{3.8\text{cm}}{384.4 \times 10^8 \text{cm}} = 0.035$ s per century, which is too large compared with the observed value of 0.019 s per century. Note that in the case of orbit of the Moon appear secular accelerations which make some deformations of the orbit, but probably they are not sufficiently large to cause some remarkable changes compared with those from the time dependent gravitational potential and the tidal accelerations.

7. Conclusion

The precise data after long time of observations of the binary pulsars yield to tests of the general relativity, specially for the formula of the gravitational radiation (3). The famous Hulse-Taylor pulsar B1913+16 is an excellent example of such confirmation. Few years ago the observations of the decay of the orbital period of the pulsar 1534+12 did not fit very precisely with the observations as it was expected, since this pulsar has a very stable timing. Now the problem seems to be overcome by introducing the theory of the linear change of the gravitational potential in the universe, which cause the increasing of the orbital period given by (28). This formula also gives good results with the PSR B1885+09, where the gravitational radiation almost vanishes. Unfortunately, formulae (3) and (28) can not be tested now for some other pulsars, because long period of observation is needed. In close future, when will be known more precise galactic data, it will be possible more rigorous tests for both formulae (3) and (28). Besides the consideration of the Hubble red shift and the increasing of the orbital period of the pulsars, the presented theory about the time dependent gravitational potential also gives some possible explanations about Pioneer anomaly acceleration and the correlations between the increasing of the average length of day, i.e.the average decreasing of the Earth spin rotation, the increasing of the lunar orbital period and the increasing of the distance from Earth to the Moon.

References

[1] Anderson, J. D.; Laing, P. A.; Lau, E. L.; Liu, A. S.; Nieto, M. M.; Turyshev, S. G. *Phys. Rev. D* 2002, 65, 082004.

[2] Anderson, J. D.; Laing, P. A.; Lau, E. L.; Liu, A. S.; Nieto, M. M.; Turyshev, S. G. *Phys. Rev. Lett.* 1998, 81, 2858-2861.

[3] Burgay, M.; D'Amico, N. D.; Possenti, A.; Manchester, R. N.; Lyne, A. G.; Joshi, B. C.; McLaughlin, M. A.; Kramer, M.; Sarkissian, J. M.; Camilo, F.; Kalogera, V.; Kim, C.; Lorimer, D. R. In *Binary Radio Pulsars*, Ed. Rasio, F. A.; Stairs, I. H.; ASP Conf. Ser. Vol. TBD 2004.

[4] Champion, D. J.; Lorimer, D. R.; McLaughlin, M. A.; Cordes, J. M.; Arzoumanian, Z.; Weisberg, J. M.; Taylor, J. H. *MNRAS* (to appear).

[5] Damour, T.; Taylor, J. H. *Astrophys. J.* 1991, 366, 501-511.

[6] Flandern, T. *Astrophys. J.* 1981, 248, 813-816.

[7] Freedman, W. L.; Turner, M. S. *Rev. of Modern Phys.* 2003, 75, 1433-1447.

[8] Groten, E. In *Highlights of Astronomy;* vol. 12, XXIV General Assembly of the IAU-2000, Mancheser, UK, 7-18 August 2000, Ed. Rickman, R. Astronomy Society of Pacific, 2000, 113-116.

[9] Kaspi, V. M.; Lyne, A. G.; Manchester, R. N.; Crawford, F.; Camilo, F.; Bell, J. F.; D'Amico, N.; Stairs, I. H.; McKay, N. P. F.; Morris, D. J.; Possenti, A. *Astrophys. J.* 2000, 543, 321-327.

[10] Kaspi, V. M.; Taylor, J. H.; Ryba, M. F. *Astrophys. J.* 1994, 428, 713-728.

[11] Kramer, M. *Astrophys. J.* 1998, 509, 856-860.

[12] Lyne, A. G.; Camilo, F.; Manchester, R. N.; Bell, J. F.; Kaspi, V. M.; D'Amico, N.; McKay, N. P. F.; Crawford, F.; Morris, D. J.; Sheppard, D. C.; Stairs, I. H. *MNRAS* 2002, 312, 698-702.

[13] Lyne, A. G.; McKenna, J. *Nature* 1989, 340, 367-369.

[14] Nice, D. J.; Sayer, R. W.; Taylor, J. H. *Astrophys. J.* 1996, 466, L87-L90.

[15] Peters, P. C.; Mathews, J. *Physical Review* 1963, 131, 435-440.

[16] Prince, T. A.; Anderson, S. B.; Kulkarni, S. R.; Wolszczan, A. *Astrophys. J.* 1991, 374, L41-L44.

[17] Ryba, M. F.; Taylor, J. H. *Astrophysical J.* 1991, 371, 739-748.

[18] Splaver, E. M.; Nice, D. J.; Arzoumanian, Z.; Camilo, F.; Lyne, A. G.; Stairs, I. H. *Astrophys. J.* (accepted for publication).

[19] Stairs, I. H.; Arzoumanian, Z.; Camilo, F.; Lyne, A. G.; Nice, D. J.; Taylor, J. H.; Thorsett, S. E.; Wolszczan, A. *Astrophysical J.* 1998, 505, 352-357.

[20] Stairs, I. H.; Thorsett, S. E.; Taylor, J. H.; Wolszczan, A. *Astrophysical J.*, 2002, 581, 501-508.

[21] Taylor, J. H.; Cordes, J. M. *Astrophys. J.* 1993, 411, 674-684.

[22] Taylor, J. H.; Weisberg, J. M. *Astrophysical J.* 1982, 253, 908-920.

[23] Thorsett, S. E.; Arzoumanian, Z.; McKinnon, M. M.; Taylor, J. H. *Astrophys. J.* 1993, 405, L29-L32.

[24] Trenčevski, K. *Gen. Rel. Grav.* 2005, 37 (3), 507-519.

[25] URL: http://sunearth.gsfc.nasa.gov/eclipse/SEhelp/ApolloLaser.html

[26] URL: http://www.astro.uu.nl/~strous/cgi-bin/glossary.cgi?l=en&o=month

[27] URL: http://www.astro.uu/nl/~strous/AA/en/antwoorden/kalenders.html

[28] Weisberg, J. M.; Taylor, J. H. In *Radio Pulsars*, Ed. Bailes, M.; Nice, D. J.; Thorsett, S. E.; *ASP Conf. Ser.* **302**, 2003, p.93.

[29] Weisberg, J. M.; Taylor, J. H. In *Binary Radio Pulsars* Ed. Rasio, F. A.; Stairs, I. H.; *ASP Conf. Ser.* Vol. TBD 2004.

[30] Will, M. C. *Theory and experiment in Gravitational Physics*, Cambridge University Press, Cambridge, US, 1993.

[31] Woods, P. T.; Shotton, K. C.; Rowley, W. R. C. *Applied Optics* 1978, 17, 1048-1054.

In: Pulsars: Discoveries, Functions and Formation
Editor: Peter A. Travelle, pp. 35-59

ISBN 978-1-61122-982-0
© 2011 Nova Science Publishers, Inc.

Chapter 3

PULSAR DISTANCES AND THE ELECTRON DISTRIBUTION IN THE GALAXY

Efe Yazgan[1]*, Oktay H. Guseinov[2] and Sevinç Tagieva[3]
[1]Middle East Technical University, Department of Physics,
Ankara 06531, Turkey
[2]Akdeniz University, Department of Physics,
Antalya, Turkey
[3]Academy of Science, Physics Institute,
Baku 370143, Azerbaijan Republic

Abstract

In this work, we present a phenomenological approach to determine pulsar distances and the average electron distribution in the Galaxy. We present the pulsar distances as a function of dispersion measure represented as graphs, not as closed mathematical formulae. The method is analogous to the widely used method which utilizes the optical absorption (A_V) to determine distances of stars in different directions. To have reliable dispersion measure - pulsar distance relations, we have used several natural requirements and distances of pulsar-calibrators. Having obtained trustworthy distances, we have constructed dispersion measure - distance relations for pulsars for 48 different regions in the Galaxy. Using the dispersion measure - distance graphs in this work, the distance to any pulsar can be determined if its dispersion measure value is known.

Key Words: Pulsar Distances, Electron Distribution, Galaxy

1. Introduction

Determining the distances of pulsars is a difficult task. Most of the pulsars detected up to date are farther than 3-4 kpc and pulsars are isolated objects. Moreover, there is no pulsar parameter which can be used to estimate the distance directly unlike the normal stars whose distances can be calculated directly from their apparent brightness. For the nearest pulsars

---

*E-mail address: efyazgan@metu.edu.tr

the parallax method can be used, however this is useful only for a few of the pulsars. For the pulsars which are genetically connected with supernova remnants, the distance values are calculated relatively accurately using the distance of the supernova remnant. But it is necessary to note that for most of these remnants distances may be estimated with an error not less than 30%. For the pulsars which are born in globular clusters the known distances of the globular clusters are used to determine the distances of pulsars. For this case the error is several times smaller than that of the supernova remnants. However, most of the pulsars are close to the Galactic plane. Thus, the pulsars in the globular clusters do not have much use in determining the distances of most of the pulsars. However, there is a direct method to determine pulsar distances through measuring the distance from the *dispersion measure* which will be explained below.

The radio pulses suffer frequency dispersion in passing through the ionized interstellar matter. The radio pulse will be dispersed by an amount of

$$\Delta t = \frac{k}{\nu^2} \int_0^d n_e dl \qquad (1)$$

where d is the distance to the pulsar and k is a known constant (see Longair 1999). The above integral is the so-called dispersion measure and denoted by DM. Stretching of the pulse can also be written using the dispersion measure as

$$\Delta t = (\frac{202}{\nu_{MHz}^3})DM \quad ms/MHz \ bandwidth \qquad (2)$$

(see Lyne & Graham-Smith 1998). Dispersion measure can be measured very accurately. We see from the above formulae that dispersion measure depends on electron distribution and the distance. Thus, fixing the dispersion measure relates electron density and distance. In this work, we give the dependence between the dispersion measure and distances for radio pulsars in each different direction, similar to the dependence between optical absorption A_V and distance. Thus when the dispersion measure is calculated from timing observations, we can find the distance using the dispersion measure - distance relations presented in this paper.

2. Electron Distribution in the Galaxy

Irregularities have been observed in the distribution of dust, molecular clouds and neutral hydrogen in the Galaxy. Considerable variations in opacity and interstellar polarization can be observed for stars with the same distance in a very small region of the sky ($\sim 1°$ square) close to the Galactic plane (Neckel & Klare 1980). This is due to the highly inhomogeneous distribution of dust clouds. For determination of the hydrogen column density along the line of sight, there are two surveys that included large number of stars; one observed 554 stars (Diplas & Savage 1994) and the other observed 594 stars (Fruscione et al. 1994). Both of these surveys show that neutral hydrogen distribution and its degree of irregularity are quite different from that of dust and molecular clouds. Irregularities in the distribution of neutral hydrogen is naturally small when compared with the dust distribution. (Ankay & Guseinov 1998). It is also natural to expect irregularities in the electron distribution.

However the degree of irregularity in the electron distribution in the Galaxy is very small when compared with that of the dust or the neutral hydrogen distribution.

The dispersion measure, which is related with the electron distribution, can be different for different pulsars with similar distances. These irregularities in the electron distribution are due to the contribution of both HII regions and supernova remnants along the line of sight. They also depend on the gravitational potential, the gas temperature and the cosmic ray distribution in the Galaxy. Even, in the halo, novae and planetary nebulae have very small contribution to the electron density in the large scale and to the background radio radiation. As indicated before the irregularities in the electron distribution is considerably smaller than the other components of the interstellar medium. Even though these irregularities are small, there is no simple model for Galactic electron distribution to calculate the distance of each pulsar because the electron distribution is neither homogeneous nor isotropic. Pulsar astronomers mostly use a mathematical model in the form of a closed formula for electron distribution to find the distances of radio pulsars in the Galaxy (e.g. Taylor & Cordes 1993 (TC93), Cordes & Lazio 2003). Using such models, distances can mostly be found with large errors since the electron distribution is not so simple to be absorbed in a closed formula. Moreover, even some of the well-known pulsar distances do not agree with the used models. Those radio pulsars are considered as exceptional and their distances are given separately from the model. These are expected from such mathematical models. But we should not expect that the situation will change considerably by improving such models with the addition of new parameters or using more complicated functions to fit with observations. This is because we do not have the necessary data on the electron distribution for large distances. We have detailed data only on the direction and distribution of large HII regions in the Galaxy and supernova remnants with large surface brightness, but not on their distances. To constrain the electron distribution model without using pulsar distances, it is necessary to know the distribution of HII regions and other objects which have considerable contribution on the electron density. Unfortunately, we have very poor data.

In this work, we determine the distances to radio pulsars by imposing natural requirements described in Section 4 and using the distances which are known precisely by independent methods. This provides us with the knowledge to construct dispersion measure-distance relations for radio pulsars. This also provides accurate information on the Galactic electron distribution.

Number density distribution of electrons in the Galaxy, when observed from the Solar system, depends on the galactic longitude(l), latitude(b) and distance from the Sun(d). However, in some directions the number density of electrons change considerably with small changes of l, b or d. This was taken into account before when estimating the distances of pulsars (Guseinov et al. 1978; TC93). Dispersion measure for pulsars naturally depends on the number of HII regions and supernova remnants on the line of sight and their electron capacities. Naturally, for pulsars with small distances from the Sun the value of the dispersion measure strongly depends on the number of HII regions and supernova remnants on the line of sight, mainly in the anticenter direction because in this case the relative contribution of these objects is large. The best example for this is the Vela pulsar. This is not expected for distant pulsars (with distances more than 3-4 kpc) especially in the Galactic center direction where the number of such objects is large. Evidently, the

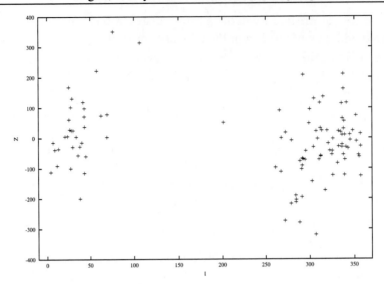

Figure 1. Galactic height (z) vs. Galactic longitude (l) for 108 pulsars with age<5 10^5 years and d>5 kpc.

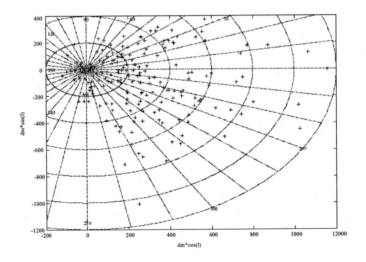

Figure 2. Dispersion measure versus Galactic longitude (l) for 189 pulsars with age<5 10^5 years.

contribution of electrons between the arms to the value of dispersion measure must be less. Figure 2 displays the dispersion measure distribution for pulsars with characteristic times $\tau \leq 5 \times 10^5$ yrs with respect to the galactic longitude l. As seen from the figure dispersion measure values are clustered around certain values of l. This is most easily observed for longitudes $l = 300°$ and $l = 360°$. Figure 3 represents the galactic distribution of pulsars with $\tau \leq 5 \times 10^5$ yrs or equivalently the distribution of giant HII regions (for the distance values see Guseinov et al. 2003a). We see that pulsars are clustered around some longitudes. We compare Figures 2 and 3 with the galactic arm structure given in Georgelin and

Georgelin (1976) and Paladini et al. (2003) which was constructed by utilizing the space distribution of giant HII regions.

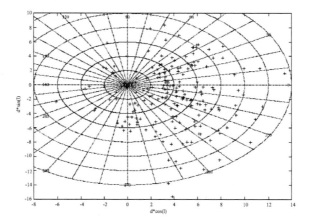

Figure 3. Galactic distribution of 189 pulsars with age<5 10^5 years.

Clusterings of young pulsars may be an evidence that the dominant number of pulsars are born inside or close to star formation regions. Star formation regions and OB associations spatially occupy only a small part of the arms, where birth rate of pulsars is large. Evidently, the electron density and the total number of electrons are considerably higher in HII regions and supernova remnants. But the value of dispersion measure, as seen from Figure 2 and 3, changes almost smoothly in the galactic plane. Therefore the value of the electron density is also large for regions between the arms in the Galactic center direction. Naturally the number density of electrons increases rapidly as we move to the galactic center direction.

In Figure 2, the considerable difference between the dispersion measure values for pulsars with approximately equal distances (see Figure 3) from the Sun in symmetrical longitude directions l about $300°$ and l about $60°$ are seen. As it is known these different directions have been scanned carefully with almost the same accuracy in pulsars surveys of the Galactic plane. The considerably lower values of dispersion measure for longitudes $\sim 60°$ and $\sim 300°$ is related with the smaller number of HII regions in the direction $l = 60°$ when compared to that of $l = 300°$ at the same distance values.

3. Progenitors, Star Formation Regions, Luminosity and Space Velocities

In the below subsections, we discuss the necessary information in the construction of dispersion measure - distance relations and in determining the electron distribution.

3.1. Progenitors of Pulsars and Star Formation Regions

Boundary mass of the progenitors of white dwarfs and neutron stars ranges from about 5 $M_\odot$(Lorimer et al. 1993) up to 6-11 $M_\odot$(Weidemann 1990). The boundary mass can be taken as $7\text{-}8M_\odot$ (Nomoto 1984; Aydin et al. 1996). This mass value corresponds to main sequence stars with spectral class B3V. Thus O and early type B stars are the progenitors of neutron stars. Error in this mass value (or equivalently the spectral class) is large for neutron stars which originate in close binary systems. According to Garmany & Stencel (1992) 75% of O stars and 58% of stars with spectral classes B2V and earlier, are born in OB associations. In addition, many young open clusters contain O and early type B stars and also massive supergiants and they are located far from OB associations. However, progenitors of pulsars may also be located in regions where there are no nearby young open clusters or OB associations. For instance, near the Sun, within a region of radius 110 pc, where there are no young open clusters (Zombeck 1990), there are six stars with masses about 9 $M_\odot$. Similarly there neither bright young open clusters nor HII regions in the vicinity of the Crab pulsar (Lynga 1987; Barbon & Hassan 1996). In some cases the progenitor of the neutron star may be a runaway star.

Size and mass of molecular clouds change in very large intervals (Patel and Putritz 1994). Generally, molecular clouds are located in the Galactic arms. But the distribution of molecular clouds are highly inhomogeneous. Therefore stars may be born in small groups (open clusters) or big clusters (OB associations). Big associations can be composed of several open clusters with different ages up to $(5-6) \times 10^7$ years. In other words, the star formation process in one large OB association can continue for such a long time (Aydın et al. 1997). As the average life-times of most of the single pulsars do not exceed this value, in (5-6) 10^7 years the positions of the Galactic arms, big star formation regions and OB associations do not change in the Galaxy considerably (Efremov 1989).

Dependence of the number of main sequence stars on their masses has an exponential character. For different star formation regions in the Galaxy, as well as in the Magellanic Clouds, the degree of the exponent may change from 1.5 up to 2.5 (Salpeter 1955, Blaha and Hamphreys 1989, Parker and Garmany 1993, Garmany et al. 1982). The uncertainty of the degree of the exponent is larger for massive stars. Evidently, the uncertainties in the exponent lead to uncertainties in number-mass distribution of pulsar progenitors. Thus we may conclude that the region giving birth to pulsars is not limited with the size of the star formation regions. This is also supported by the observations of young pulsars and supernova remnants which are closer than 3 kpc to the Solar system. These observations show that these objects might be originated far from star formation regions and OB associations. This is because the distribution of young pulsars and supernova remnants in the neighborhood of the Sun closer than 3 kpc indicate that these objects may be born far from star formation regions and OB associations.

The distance of radio pulsars and the young objects such as HII and HI regions and molecular clouds, on the average, cannot be determined with a precision better than 30%. Therefore the location of star formation regions for which we have not detected enough number of massive stars, can only be defined roughly. Despite that, their directions are very well known.

3.2. The Radio Luminosity of Pulsars

The luminosity of single radio pulsars practically does not depend on the spin period (P) and period derivative ($\dot{P}$), i.e. the positions of pulsars on the P-$\dot{P}$ diagram. This shows that the radio luminosity of single radio pulsars depends very weakly on the magnetic field strength of neutron stars since the magnetic field is proportional to $\sqrt{P\dot{P}}$. Moreover it does not depend on the rate of rotational energy loss either since it is proportional to $\dot{P}/P^3$ (Allakhverdiev & Tagieva 2003). Therefore we may conclude that the radio luminosity of single pulsars practically does not depend on the age, mass and duplicity of their progenitors. However it is important to note that this is inapplicable for close binary systems.

3.3.   Space Velocity of Pulsars

The average velocity of single radio pulsars is about 250 km/s (Allakhverdiev et al. 1997; Hansen & Phinney 1997). Therefore radio pulsars with ages up to 5×10^5 years can not move far from their birth places to distances more than 200 pc (which corresponds to 160 pc if we project this velocity to the Galactic plane). But even for pulsars whose velocities are very high, it is practically impossible for them to pass from one arm to another in a time shorter than 10^6 years. It is also necessary to note that the high velocity pulsars are very rare. Size of big star formation regions which are located in the arms may be as large as 1 kpc. The distance between the arms is at least 1 kpc. Therefore pulsars with ages up to $\sim 5 \times 10^5$ years can be confidently considered to be at their birth places.

3.4.   Deviation of Star Formation Regions from the Galactic Plane

Only old population stars (i.e. stars with ages about 10^{10} years) can achieve dynamical stability. The scale height and number density distributions in the Galactic center direction of old and young population stars are different. This shows that old and young populations are not in equilibrium with each other. The total mass of gas and young stars located in the Galactic arms is only 1-3% of the mass of the whole Galaxy and the parameters of these arms change with a characteristic time of $\sim10^9$ years. Star formation regions in the arms are more unstable having ages about an order of magnitude less than the characteristic ages of the arms of the Galaxy. The arms are neither in dynamical equilibrium with different populations of the Galaxy, nor with each star formation region. Therefore for some of the regions, the arms (star formation regions) do not coincide with the geometrical plane of the Galaxy. For some regions star formation regions may be higher or lower from the galactic plane (Efremov 1989). As shown by optical observations of cepheids with high luminosities (with long pulsation periods) and red supergiants, which are located at distances about 5-10 kpc from the Sun, in the longitude interval $l = 220° - 330°$, star formation regions are located about 300 pc below the plane of the Galaxy. At the same distance, in the direction $l = 70° - 100°$ it is found that star formation regions are placed about 400 pc above the plane of the Galaxy. The young objects closer than 3-5 kpc in the direction $l = 270° - 320°$ are located about 150 pc down the geometric plane of the Galaxy (Berdnikov 1987). Figure 1 displays longitude(l) versus galactic height(z) of 108 radio pulsars with characteristic age $\tau < 5\times10^5$ years and distances > 5 kpc. As seen from the figure, between $250° < l < 300°$ young pulsars are below the galactic plane and between $50° < l < 80°$ they are above the plane. These deviations are taken into account in estimating distances of radio pulsars and in the construction of dispersion measure - distance relations.

4.   Preparing the Pulsar Distance Sample

In order to investigate the arm structure and the locations of star formation regions around the Sun within a distance of 3–4 kpc, usually OB associations, open clusters and red supergiants are studied. For these objects the relative errors in estimating their distances is about 30% (Humphreys 1978; Efremov 1989; Aydın et al. 1996; Ahumada & Lapasset 1995). It is difficult to estimate the distance of all the extended objects belonging to the arms (molecular and HI clouds, supernova remnants and HII regions). In determining the distances to these objects using neutral hydrogen 21 cm line and Galaxy rotation models, the error is more than 30%. It is known that it is not possible to calculate the distance to an object using neutral hydrogen line at 21 cm if the galactic radial velocity component of the object is small. In addition to this, for certain longitudes (close to the Galactic center direction) and distances the hydrogen lines gives two or three different distance values for distant or extended objects. Despite that, it is the most widely used model. But for the current work, in determining the distance calibrators for pulsars, distance estimates calculated using the 21 cm line are not used.

It is a well known fact that no relation has been found in pulsar parameters to estimate their distance unlike the normal stars. In the early days of pulsar astronomy between 1970 and 1980, the data on the origin of pulsars, the mass of their progenitors and their birth rates were poor. Therefore, a homogeneous electron density distribution in the Galaxy had been assumed. Star formation regions, HII regions and supernova remnants is mainly in the Galactic arms and arm structures are known. One should also know some of the pulsar distances independent of their dispersion measure to be used as calibrators. Thus, pulsars with distances estimated using HI line in general had been used as an extra distance calibrator (Manchester & Taylor 1981). However, later on pulsar distances have been estimated according to the rough model of Galactic electron distribution and some natural requirements (Guseinov & Kasumov 1981; Johnston et al. 1992; TC93; Gök et al. 1996). Today, calibrators are chosen from members of globular clusters or Magellanic clouds, pulsars connected to supernova remnants with known distances and from pulsars whose distances are known from other trustworthy data (Lyne & Graham-Smith 1998). For the current work, in determining the calibrators for pulsars, distance estimates calculated using the 21 cm line are not used.

When estimating pulsar distances, the model of Galactic electron distribution developed by TC93 has been widely used in recent years. In estimating the pulsar distances, the approach of Gök et al. (1996) gave smaller distances than the ones calculated using the model of TC93 for pulsars farther than 4 kpc and also for pulsars with Galactic latitudes greater than about 10° at distances d>1 kpc. The new model of electron distribution (Cordes and Lazio 2003) predict distances close to the values given by Gök et al. (1996). Gómez et al. (2001) have published a huge pulsar list which could be used as calibrators in the formation of a new model of electron distribution. We have revised their distance values to use them as distance calibrators. In Table 1 we present 39 pulsars for which the errors in the distance values are not more than 30%. Since the distances of pulsars in the same globular cluster are the same, only one pulsar from each globular cluster has been included in the table. Number of pulsars which we used as calibrators, is considerably smaller than the one in the calibrator list of Gómez et al. (2001). In this table, one of the most problematic calibrators is PSR J0835-4510 (in Vela supernova remnant). The distance to the Vela remnant is given as d=250 pc (Ogelman et al. 1989; Cha et al. 1999; Danks 2000) and $d \approx 280$ pc (Bocchino et al. 1999). But we have to take into account the fact that Vela supernova remnant is expanding in a dense medium. The magnetic field is given as 6×10^{-5} Gauss (de Jager et al. 1996) and as $(5-8.5) \times 10^{-5}$ Gauss (Bocchino et al. 2000) and supernova explosion energy is $(1-2) \times 10^{51}$ ergs (Danks 2000). These values are more than average for a pulsar. Evidently, the uncertainties are high for these quantities. However, if we assume that these data are correct and the distance to the Vela remnant is 250 pc, then in the $\Sigma - D$ diagram Vela and the SNR G 327.6+14.6 are found to be located at the same position. However, this is not convincing, since Vela and SNR G 327.6+14.6 are formed from different type of supernova explosions, i.e. SNR G 327.6+14.6 was formed from a type Ia explosion (Hamilton et al. 1997) whereas Vela was formed from a type II explosion. This SNR has a distance of 500 pc from the Galactic plane, and so it lies in a much less dense medium than the medium of the Vela remnant. In the $\Sigma - D$ diagram it is hard to understand the deviation of the Vela remnant from the $\Sigma - D$ relation. On the other hand, the deviation of SNR G 327.6+14.6 is readily understandable, because this supernova remnant is formed from a type Ia explosion, but the $\Sigma - D$ relation is constructed for S and C type supernova remnants. The distance of OB associations including the young open clusters in the direction of Vela remnant are not closer than 0.25 kpc. One of these open clusters which lie in the OB association (OB2) closer to the Sun, in the direction of Vela, is Pismis 4 ($l = 262.7°$, $b = -2.4°$). Pismis 4 has a well known distance of 0.6 kpc (Ahumada & Lapasset 1995; Aydin et al. 1997). Since the progenitors of supernova remnants (or pulsars) are massive stars, the probability that Vela is not at 0.25 kpc but closer to a star formation region is high. PSR J1302-6350 ($l = 304.2°$, $b = -0.9°$) has the second highest electron density (after Vela) which is 0.113 cm^{-3}. This pulsar has a Be type companion having a variable wind. The pulsar is at a distance of 1.3 kpc (Johnston et al. 1994). So, it is not unexpected for this pulsar to have such a high electron density. PSR J1644-4559 ($l = 339.19°$, $b = -0.19°$; d=4.5 kpc) has the third highest electron density with a value of 0.107 cm^{-3}. Since the luminosity at 1400 MHz for this pulsar is higher than all other pulsars with known flux ($L_{1400} = 6.3 \times 10^3$, $L_{400} = 7.9 \times 10^3$), it would not be true to adopt its distance to be more than 4.5 kpc. PSR J1341-6220 ($l = 308.7°$, $b = -0.4°$) has an average electron density of 0.091 cm^{-3} at 8 kpc. For SNR G308.8-0.1 which is thought to be genetically connected with this pulsar, the angular size and flux values are not precisely known. But we can not accept that this supernova remnant is located farther. Due to these reasons the distance of the pulsar-supernova remnant pair must be about 8 kpc (using the $\Sigma - D$ relation). In the direction and at the distance of this pulsar, star formation region lies below the Galactic plane. Consequently, electron density shows asymmetric distribution about the plane, and the value of the electron density below is higher than the corresponding electron density at the same longitude with the same latitude but above the plane. For all of the pulsar sample, average value of electron density is around 0.04 cm^{-3}. Assuming 250 pc for the distance of Vela leads to an electron density two times higher than the electron density calculated assuming the distance 400 pc. However, even when the distance is assumed to be 400 pc, electron density is readily the highest of all. So distance of Vela is at least 0.4 kpc and the most recent supernova catalog (Green 2001) gives the distance of Vela as ~ 0.5 kpc. The value of the electron density along the line of sight is high for Vela pulsar because it lies in a supernova remnant in the nearest OB association. There are about 15 pulsars genetically connected to supernova remnants in our Galaxy (Kaspi and Helfand 2002). Some of these supernova remnants most probably are in OB associations. However, since these associations are distant sources, the column density is high and the value of the electron density along the line of sight is lower.

Table 1. Distance calibrators

Name	l	b	d	DM	n_e	Location	Ref
0024-7204W	305.9	-44.9	4.5	24.3	0.005	GC NGC 104 (47 Tuc)	[1,2]
0045-7319	303.5	-43.8	57	105.4	0.002	SMC	[3]
0113-7220	300.6	-44.7	57	125	0.002	SMC	[4]
0205+6449	130.7	3.1	3.2	140.7	0.044	SNR G130.7+3.1	[5]
0455-6951	281.2	-35.2	50	94.9	0.002	LMC	[4]
0529-6652	277.0	-32.8	50	103.2	0.002	LMC	[4]
0534+2200	184.6	-5.8	2	56.79	0.028	SNR G184.6-5.8(Crab)	[6]
0535-6935	280.1	-31.9	50	89.4	0.002	LMC	[4]
0540-6919	279.7	-31.5	50	146	0.003	LMC	[7]
0826+2637	196.9	31.7	0.4	19.48	0.049	Parallax	[8]
0835-4510	263.6	-2.8	0.40	68.2	0.152	SNR G263.9-3.3(Vela)	[9,10,11]
0922+0638	225.4	36.4	1.21	27.31	0.020	Parallax	[12,13]
0953+0755	228.9	43.7	0.28	2.97	0.011	Parallax	[8,14]
1119-6127	292.2	-0.54	10	707.4	0.101	SNR G292.2-0.5	[15,11,16]
1124-5916	292.0	1.8	6	330	0.066	SNR G292.0+1.8	[17]
1302-6350	304.2	-0.9	1.3	146.72	0.113	Sp Binary, Be	[18]
1312+1810	332.9	79.8	18.9	24	0.001	GC NGC 5024 (M53)	[1,19]
1341-6220	308.7	-0.4	8.5	717	0.097	SNR G308.8-0.1	[20,11]
1456-6843	313.9	-8.5	0.45	8.6	0.019	Parallax	[21]
1513-5908	320.3	-1.2	4.2	253.2	0.060	SNR G320.4-1.2	[7,11,16]
1518+0204B	3.9	46.8	7	30.5	0.004	GC NGC 5904 (M5)	[1,22,23]
1623-2631	350.9	15.9	1.8	62.86	0.035	GC NGC 6121 (M4)	[1,24]
1641+3627B	59.8	40.9	7.7	29.5	0.004	GC NGC 6205 (M13)	[1,25]
1701-30	353.6	7.3	5	114.4	0.023	GC NGC 6266 (M62)	[1,26,27]
1721-1936	4.9	9.7	9	71	0.008	GC NGC 6342	[1,28]
1740-53	338.2	-11.9	2.3	71.8	0.031	GC NGC 6397	[1,26,29]
1748-2021	7.7	3.8	6.6	220	0.033	GC NGC 6440	[1,30]
1748-2445B	3.85	1.7	5	205	0.029	Ter 5	[1,31]
1801-2451	5.27	-0.9	4.5	289	0.064	SNR G5.27-0.9	[32]
1803-2137	8.4	0.1	3.5	233.9	0.067	SNR G8.7-0.1	[33,34,35,11,16]
1804-0735	20.8	6.8	7	186.38	0.027	GC NGC 6539	[1,36]
1807-2459	5.8	-2.2	2.6	134	0.052	GC NGC 6544	[1,26,37]
1823-3021B	2.8	-7.9	8	87	0.011	GC NGC 6624	[1,38]
1824-2452	7.8	-5.6	5.7	119.83	0.021	GC NGC 6626 (M28)	[1,39]

Radio luminosity of pulsars can not depend on their birth place and can not considerably exceed the luminosities of the most luminous pulsars (e.g. The most luminous pulsar in the Magellanic clouds and the Crab pulsar whose distances are very well known). Luminosity

Table 1. Continued

Name	l	b	d	DM	n_e	Location	Ref
1856+0113	34.6	-0.5	2.8	96.7	0.035	SNR G34.7-0.4	[11,16]
1910-59	336.5	-25.6	4	34	0.009	GC NGC 6752	[1,26,40]
1910+0004	35.2	-4.2	6.5	201.5	0.031	GC NGC 6760	[1,28]
1930+1852	54.1	0.27	6	308	0.062	SNR G54.1+0.3	[41]
1932+1059	47.4	-3.9	0.17	3.18	0.019	Parallax	[42,43,44,45]
1952+3252	68.8	2.8	2	44.98	0.022	SNR G69.0+2.7	[7,34,11,16]
2022+5154	87.9	8.4	1.1	22.58	0.021	Parallax	[46]
2129+1209H	65.1	-27.3	10	67.15	0.007	GC NGC 7078 (M15)	[7]
2337+6151	114.3	0.2	2.5	58.38	0.021	SNR G114.3+0.3	[47,11,16]

[1] Harris 1996; [2] Hesser et al. 1987; [3] Feast & Walker 1987; [4] Crawford et al. 2001a; [5] Camilo et al. 2002b; [6] Trimble & Woltjer 1971; [7] Taylor et al. 1996; [8] Gwinn et al. 1986; [9] Cha et al. 1999; [10] Legge 2000; [11] Guseinov & Ankay 2002; [12] Chatterjee et al. 2001; [13] Fomalont et al. 1999; [14] Brisken et al. 2000; [15] Crawford et al. 2001b; [16] Kaspi & Helfand 2002; [17] Camilo et al. 2002a; [18] Johnston et al. 1994; [19] Rey et al. 1998; [20] Caswell wt al. 1992; [21] Bailes et al. 1990; [22] Brocato et al. 1996a; [23] Sandquist et al. 1996; [24] Cudworth & Rees 1990; [25] Paltrinieri et al. 1998; [26] D'Amico et al. 2001; [27] Brocato et al. 1996b; [28] Heitsch & Richtler 1999; [29] Alcaino et al. 1987; [30] Ortolani et al. 1994; [31] Ortolani et al. 1996; [32] Thorsett et al. 2002; [33] Frail et al. 1994; [34] Allakhverdiev et al. 1997; [35] Finley & Ögelman 1994; [36] Armandroff 1988; [37] Kaspi et al. 1994; [38] arejedini & Norris 1994; [39] Rees & Cudworth 1991; [40] Buonanno et al. 1986; [41] Camilo et al. 2002c; [42] Weisberg et al. 1980; [43] Salter et al. 1979; [44] Backer & Sramek 1982; [45] Campbell 1995; [46] Campbell et al. 1996; [47] Furst et al. 1993

of Crab pulsar is 2.6×10^3 mJy kpc^2, 56 mJy kpc^2 at 400 MHz and 1400 MHz, respectively. Luminosity of the strongest pulsar in the Large Magellanic Cloud (PSR J0529-6652) is 6×10^3 mJy kpc^2 at 400 MHz and 7.5×10^2 mJy kpc^2 at 1400 MHz. PSR J0045-7319 which is in the Small Magellanic Cloud has the highest luminosity at 400 MHz. Its luminosity at this frequency is 9.5×10^3 mJy kpc^2 whereas its luminosity at 1400 MHz is 1.3×10^3 mJy kpc^2. Therefore the upper limit for luminosities of pulsars might be close to the value of 1.5×10^4 mJy kpc^2 and 3.5×10^3 mJy kpc^2 for 400 MHz and 1400 MHz, respectively (spectral indices of radiation in radio band of pulsars have been also taken into account). In our list of whole pulsars the strongest one among the galactic pulsars is PSR J1644-4559 ($l = 339.19$, $b = -0.19$; $d = 4.5$ kpc)with luminosities of 7.6×10^3 mJy kpc^2 and 6.3×10^3 mJy kpc^2 at 400 and 1400 MHz, respectively (Guseinov et al. 2003b and references therein). These pulsars have extraordinarily flat spectra (Taylor et al. 1996).

Since pulsars were born in the Galactic plane and surveys have scanned the Galactic plane many times, most of the known pulsars, especially the farthest ones, have small Galactic latitudes ($|b| < 5°$) (Manchester et al. 2002; Morris et al. 2002, Lyne et al. 1998; D'Amico et al. 2001; Manchester et al. 1996). As can be seen in our calibrator table (Table 1), there are 13 pulsars with $|b| > 30°$, 6 in Magellanic Clouds, 4 in globular clusters and 3 with known parallax, 10 pulsars with $30° > |b| > 7°$ from which 8 of them belongs to globular clusters and 2 have parallax, 7 pulsars with $7° > |b| > 3°$ from which 4 are located in globular clusters and 1 have known parallax measurement, and only 14 pulsars with $|b| < 3°$. Thus, the calibrators in Magellanic Clouds, globular clusters and calibrators with known trigonometric parallax's become insignificant for pulsars with small $|b|$. Only 3 from our calibrator list belong to $|b| < 3°$ and have distances greater than 5 kpc. Therefore, for the pulsars with large distances and low $|b|$ values, in other words for Galactic field pulsars, there are almost no calibrators. In addition to this, for such distances it is quite difficult to judge the value of the electron density. Because we also do not have the necessary data on the star formation regions, HII regions and supernova remnants and their electron capacity. It is also not easy to find pulsar distances (estimations of the electron distribution) for pulsars with $d < 5$ kpc without the use of natural criteria.

Considering the discussions above in adopting distances for pulsars, the following criteria become very important:

1. In the direction of $40° < l < 320°$ we see the most luminous pulsars throughout the Galaxy.

2. For all Galactic longitudes (l), pulsars with equal characteristic times (τ) must have, on the average, similar $|z|$ values except for pulsars with $\tau \leq 5 \times 10^6$ years in the regions where star formation regions are considerably above or below the Galactic plane.

3. Pulsars with $\tau < 5 \times 10^5$ years must still be closer to their birth places.

4. The radio pulsar luminosity does not depend on l, d, P ot $\dot{P}$ and it should not considerably exceed the luminosity of known strongest pulsars at 400 and 1400 MHz. The value of luminosities at 400 and 1400 MHz must at most be 1.5×10^4 mJy kpc^2 and 3×10^3 mJy kpc^2 respectively. If the luminosity of a pulsar is estimated to be greater than these values, then it means that, most probably, its distance is overestimated and its distance value should be reduced.

5. Electron density in the Galaxy must be correlated with the number density of HII regions and OB associations but not strongly at large distances. Electron density must increase as one approaches to the galactic center and galactic plane similar to the case of background radiation in the radio band.

6. Pulsar distances must be arranged in such a way that their value should correspond to a suitable distance in the set of values in Table 1 (value of dispersion measure and the direction of the pulsar have to be taken into account).

7. The irregularities of the electron distribution must decrease as we move away from the Galactic plane.

5. Dispersion Measure - Distance Relations

We have adopted distance for each pulsar in such a way that pulsar distances satisfy the criteria above. We gathered the pulsar distances in different directions for different longitude

(l) and latitude (b) intervals. We have divided the Galaxy into 14 longitude intervals. Moreover, for each longitude interval we have divided pulsars into different latitude intervals. The boundaries of each interval are determined in such a way that in the selected region the dispersion measure - distance relation is almost linear. The corresponding regions are given in Figures 9a-25c. For example Figures 9a-9d represent distance (d) versus dispersion measure (DM) for pulsars with $-10° < l < 10°$ and $-3° < b < 3°$; $-7° < b < -3°$ and $3° < b < 7°$; $-15° < b < -7°$ and $7° < b < 16°$; $-25° < b < -15°$ and $15° < b < 25°$. There are totally 48 regions. The number of such regions might increase in the future as more pulsars are discovered and more refined relations can be obtained. From the slope of $d - DM$ relation in each figure, it is seen that electron density depends also on the distance from the Sun in each interval of l and b. The calibrator pulsars are displayed with their names in the figures. Our adopted distances for all 1328 pulsars, which we have used in the 48 figures in this paper are included in the list of Guseinov et al. (2003a). These figures, when the dispersion measure is known give the distance to the pulsar. For instance, Figure 9b corresponds to $-10° < l < 10°$ and $3° < |b| < 7°$. If pulsar has a $|b|$ value close to $7°$ one must choose the value of the distance close to maximum among the distances which correspond to the value of DM. Evidently, if $|b|$ is close to $3°$, then we must choose the lower distance value.

In Figure 4 we compare the distances we have determined and the distances determined by TC93. For pulsars closer to the Sun, the distances we have predicted and those determined by the TC93 model are similar in general. However, for few pulsars, namely PSR J1302-6350, J1748-2021, J1804-0735 and J1910-59, we have estimated considerably different values compared to the TC93 model. Among the pulsars which are close to the Sun, PSR J1939+2134 ($l = 57.51°$, $b = -0.29°$) has the greatest difference. It is also important to note that for farther pulsars the difference between the predictions of the models differ up to more than twice. Larger values of differences exist for pulsars with higher latitudes ($|b|$) and for pulsars which are located close to the plane of the galaxy, but have large distances (Gök et al. 1996). However, most of the pulsars with our distance predictions are larger than 6-7 kpc are located in the south hemi-sphere and were discovered during the 1400 MHz survey (with l between 300°-360° and $|b|$ in most cases smaller than 2°).

In finding the pulsar distance one of the important criteria is that the pulsars at the same age should be at almost the same distance from the Galactic plane. In Figures 5 and 6, the distance from the Galactic plane (z) versus the distance from the Sun (d) for pulsars with age<10^6 years in the longitude interval $l = 300° - 360°$ is given according to our model and TC93 model of electron distribution, respectively. The cone corresponds to 1°. It gives the limits of z at each distance d. As seen from these figures there is not a considerable difference neither in z nor in d. For only four of these pulsars TC93 model has given very large distance values. This indicates that for small b values ($|b| < 2°$) and distances up to 11-13 kpc there is no significant difference in electron distributions of the two models for l between $300° - 360°$. This is in general also true for other longitude directions.

In Figures 7 and 8, the distance from Galactic plane z versus the distance from the Sun is given according to our model and TC93 model for pulsars with characteristic age τ between $10^6 - 10^7$ years. As seen from these figures, for pulsars farther than 12-13 kpc, our distance estimates (d,z) are considerably smaller than the distance estimates (d,z) given by the TC93 model. It is seen from Figure 7 that for all distances number density of pulsars are higher

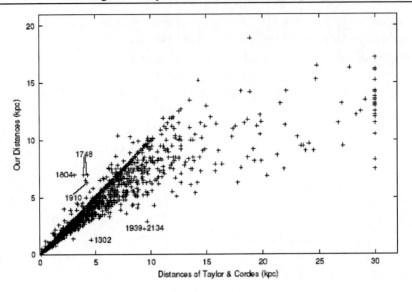

Figure 4. Our distances vs. distances predicted by TC93 model for 1318 pulsars.

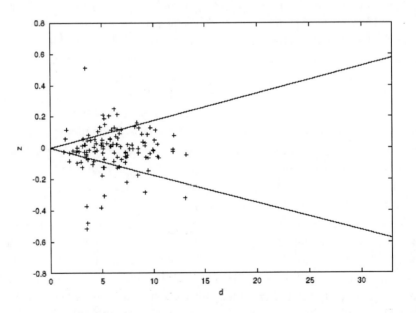

Figure 5. Our distance values vs. Galactic height (z) for pulsars with l=300°- 360°and $\tau \leq$ 10^6 years. The angle between the lines corresponds to $|b| = 1°$.

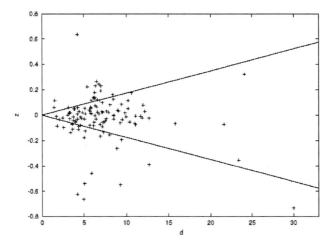

Figure 6. Distances calculated from TC93 model vs. Galactic height (z) for the same pulsars within l=300°-360° and $\tau \leq 10^6$ years. The angle between the lines corresponds to $|b| = 1°$.

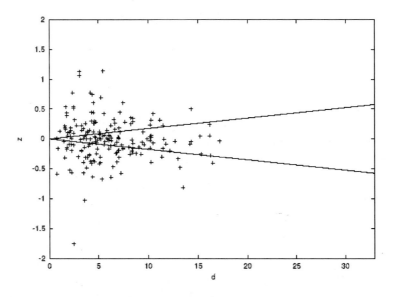

Figure 7. Our distance values vs. Galactic height (z) for pulsars with l=300°- 360° and $\tau = 10^6$ years. The angle between the lines corresponds to $|b| = 1°$.

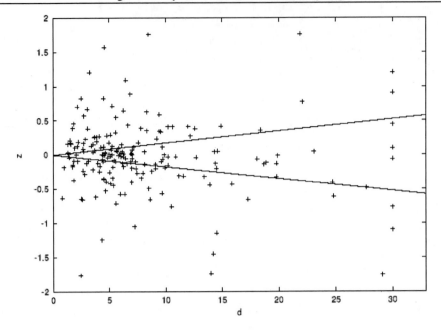

Figure 8. Distance values from TC93 model for pulsars with l=300°-360° which have ages of $10^6 - 10^7$ years.

near the Galactic plane, and for pulsars farther than 6-7 kpc the distances from the Galactic plane on the average increase slowly. The distance distribution according to the TC93 model does not agree with this and predict considerably larger distances. Thus according to the TC93 model, as the distances of pulsars increase, the average distance from the Galactic plane increases. It shows that in the TC93 model the space velocities of farther pulsars are larger. However there is no reason for that to occur. The gradual increase of distances of pulsars from the plane of the Galaxy in Figures 5 and 7 have natural explanations. On the average for all directions and distances from the Sun, the number density of electrons have a maximum value on the plane of the Galaxy. On the other hand for larger values of dispersion measures it is difficult to observe pulsars with small periods. Therefore we can not detect the young pulsars in the survey, because luminosities of pulsars practically do not depend on their location on the $P - \dot{P}$ diagram. From Figures 5-8 it can be concluded that the possibility of observation of distant pulsars are higher for older high period ones.

6.   Conclusion

In this work, we have divided the Galaxy into 48 regions for different longitude and latitude intervals. These regions are small enough to have reliable dispersion measure - distance relations. Number of regions will increase and $DM - d$ relations will improve with the accumulation of new pulsar data. These relations we have presented provide accurate distance values for pulsars whose dispersion measures are known.

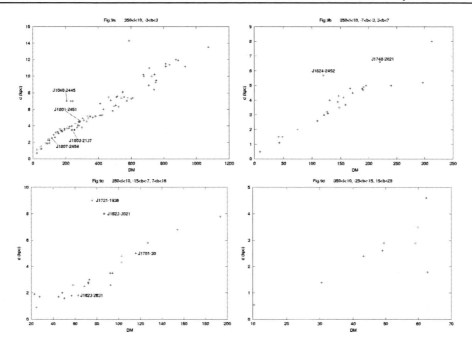

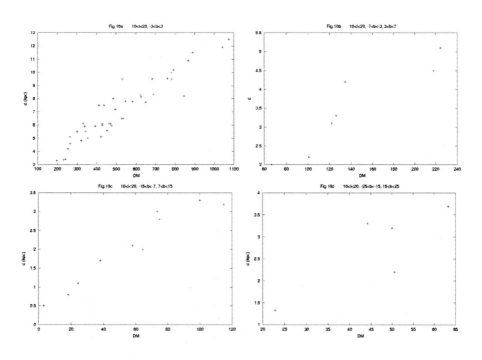

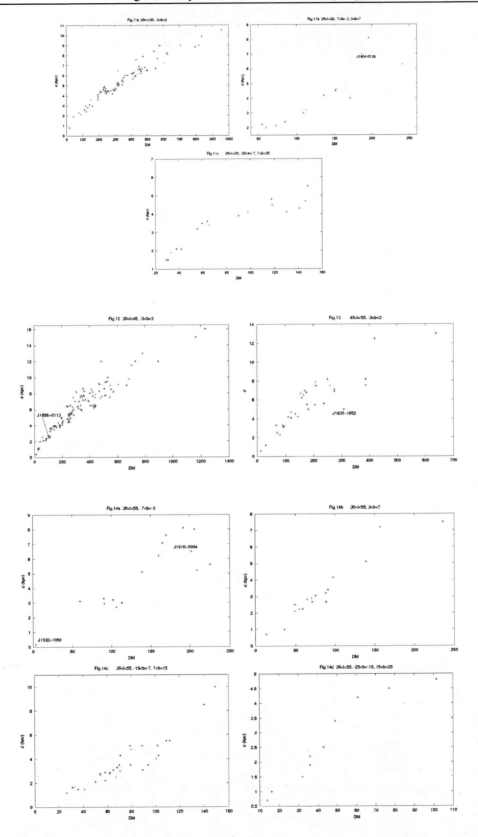

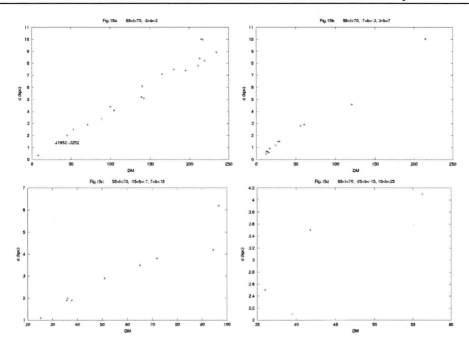

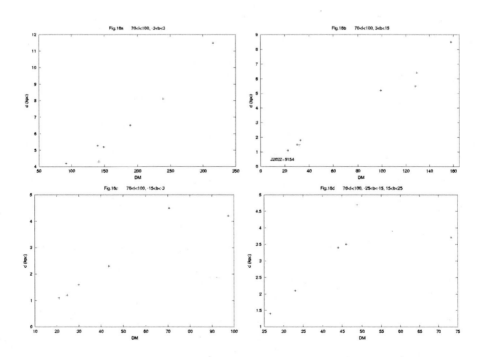

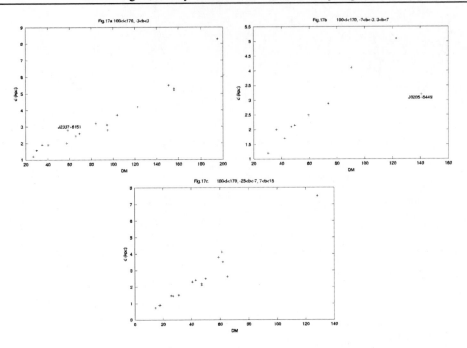

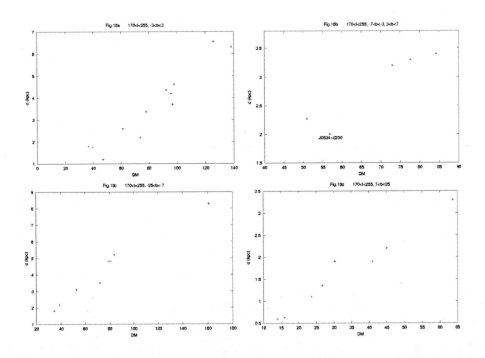

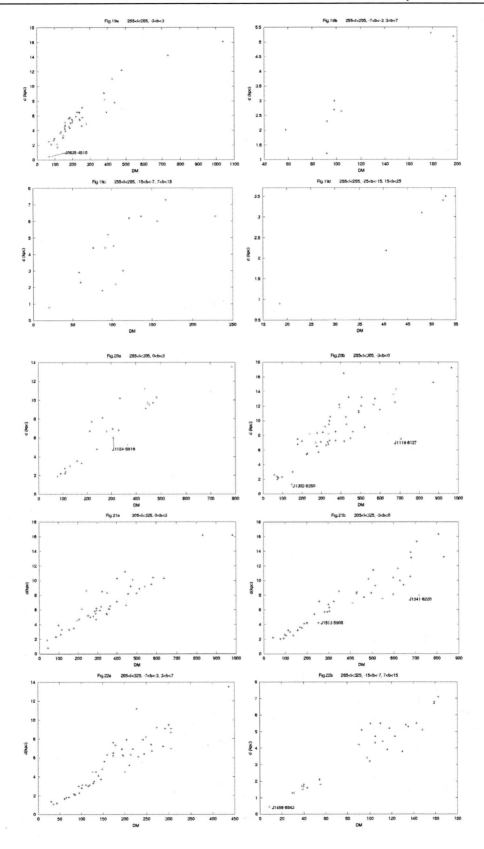

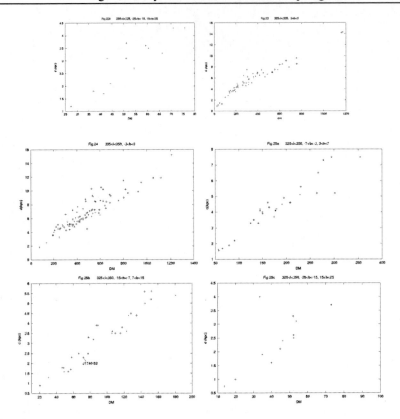

Acknowledgments

Authors would like to thank A. Kupcu-Yoldas for her help in preparing Figures 1-3.

References

Ahumada, J. and Lapasset, E., *Astron. & Astrophys. Suppl.* **109**, 375 (1995)

Alcaino, G., Buananno, R., Caloi, V. et al., *Astron. J.* **94**, 917 (1987)

Allakhverdiev, A. O., Guseinov, O. H., Tagieva, S. O., Yusifov, I. M., *Astron. Rep.* **41**, 257 (1997)

Allakhverdiev, A. O., Tagieva, S. O., *Astron. & Astroph. Trans.*, *in press* (2003)

Ankay, A. and Guseinov, O. H., *Astron. & Astrophys. Trans.* **17**, 301 (1998)

Armandroff, T.E., *Astron J.* **96**, 588 (1988)

Aydın, C., Özdemir, S., Guseinov, O. H. et al., *Turk. J. Phys.* **20**, 1149 (1996)

Aydın, C., Albayrak, B., Gözel, İ., Guseinov, O.H., *Turk. J. Phys.*, **21 N7**, 875 (1997)

Bailes, M., Reynolds, J.E., Manchester, R.N., Norris, R.P., and Kesteven, M.J. *Nature*, **343**, 240 (1990)

Backer, D.C., and Sramek, R.A. *Astrophys. J.* **260**, 512 (1982)

Barbon, R. & Hassan, S. H., *Astron. & Astrophys. Suppl.* **115**, 325 (1996)

Berdnikov, L. N., *Sov. Astron. Lett.* **13**, 45 (1987)

Blaha, C. & Hamphreys, R., *Astron. J.* **98**, 1598 (1989)

Bocchino, F., Maggio, A., Sciortino, S., *Astron. & Astrophys.* **342**, 839 (1999)

Bocchino, F., Maggio, A., Sciortino, S. and Raymond, J., *Astron. & Astrophys.* **359**, 316 (2000)

Brisken, W.F., Benson, J.M., Beasley, A.J., Fomalont, E.B., Goss, W.M., and Thorsett, S.E., *Astrophys. J.* **541**, 959 (2000)

Brocato, E., Buonanno, R., Malakhova, Y., and Piersimoni, A.M., *Astron. & Astrophys.* **311**, 778 (1996a)

Brocato, E., Castellani, V., and Ripepi, V., *Astron. J.* **111**, 809 (1996b)

Buonanno, R., Caloi, V., Castellani, V., et al., *Astron. & Astrophys. Suppl.* **66**, 79 (1986)

Camilo, F., Manchester, R.N., Gaensler, B.M., Lorimer, D.R. and Sarkissian, J., *Astrophys. J. Lett.* **567**, 71 (2002a)

Camilo, F., Stairs, I.H., Lorimer, D.R., et al., *Astrophys. J. Lett.* **571**, 41 (2002b)

Campbell, R.M., *Astronomical Distances Through Vlbi: Pulsars and Gravitational Lenses, Ph.D. Thesis.* (1995)

Campbell, R.M., N. Bartel, I.I. Shapiro et al., *Astrophys. J. Lett.* **461**, 95 (1996)

Caswell, J.L., Kesteven, M.J., Stewart, R.T., Milne, D.K., and Haynes, R.F., *Astrophys. J. Lett.* **399**, 151 (1992)

Cha, A.N., Sembach, K.R., Danks, A.C., *Astrophys. J.* **515**, L25 (1999)

Chatterjee, S., Cordes, J.M., Lazio, T.J.W. et al. *Astrophys. J.* **550**, 287 (2001)

Cordes, J.M., and Lazio, T.J.W., astro-ph/0301598 (2003)

Crawford, F., Gaensler, B.M., Kaspi, R.N. et al. *Astrophys. J.* **554**, 152 (2001)

Cudworth, K.M., and Rees, R., *Astron. J.* **99**, 1491 (1990)

D'Amico, N.,Stappers, B. W., Bailes, M. et al., *Mon. Not. Roy. Astron. Soc.* **297**, 28 (1998)

D'amico, N., Lyne, A. G., Manchester, R. N., et al., *Astrophys. J. Lett.* **548**, 2001, 171 (2001)

Danks, A.C., *Astrophys. & Spa. Sci.* **272**, 127 (2000)

De Jager, O.C., Harding, A.K., Strickman, M.S. *Astrophys. J.* **460**, 729 (1996)

Diplas, A. and Savage, B. D., *Astrophys. J. Suppl.* **93**, 211 (1994)

Efremov, Yu. N. *Sites of Star Formation in Galaxies* Moscow: Nauka. (1989)

Fomalont, E.B., Goss, W.M., Beasley, A.J. and Chatterjee, S. *Astron. J.* **117**, 3025 (1999)

Feast, M.W., and Walker, A.R., *ARA&A* **25**, 345 (1987)

Finley, J.P., and Oegelman, H., *Astrophys. J. Lett.* **434**, 25 (1994)

Frail, D.A., Kassim, N.E., and Weiler, K.W. *Astron. J.* **107**, 1120 (1994)

Fruscione, A., Hawkins, I., Jelinsky, P. and Wiercigroch, A., *Astrophys. J. Suppl.* **94**, 127 (1994)

Furst, E., Reich, W., and Seiradakis, J.H., *Astron. & Astrophys.* **276**, 470 (1993)

Garmany, C. D. et al., *Astrophys. J.* **263**, 777 (1982)

Garmany, C. D. & Stencel, R. E., *Astron. & Astrophys. Suppl.* **94**, 211 (1992)

Georgelin, Y. M. and Georgelin, Y. P., *Astron. & Astrophys.* **49**, 57 (1976)

Gómez, G. C., Benjamin, R. A. and Cox, D. P., *Astron. J.* **122**, 908 (2001)

Gök, F., Alpar, M. A., Guseinov, O. H. and Yusifov, I. M., *Turk. J. Phys.* **20**, 275 (1996)

Green, D.A., *A Catalogue of Galactic Supernova Remnants 2001 December version*, http://www.mrao.cam.ac.uk/surveys/snrs/ (2001)

Guseinov, O.H. Ankay, A., Sezer, A. and Tagieva, S., *unpublished work*, (2002)

Guseinov, O. H., Kasumov, F. K. & Yusifov, I. M., *Astrophys. & Spa. Sci.* **59**, 285 (1978)

Guseinov, O. H. and Kasumov, F. K., *Astron. Zh.* **58**, 996 (1981)

Guseinov, O. H., Yerli, S. K., Özkan, S., Sezer, A., & Tagieva, S. O., *Astron. & Astrophys. Trans.* **23**, 357 (2004)

Gwinn, C.R., Taylor, J.H., Weisberg, J.M., and Rawley, L.A., *Astron. J.* **91**, 338 (1986)

Harris, W.E., *Astron. J.* **112**, 1487 (1996)

Hamilton, A.J.S., Fesen, R.A., Wu, C.-C. et al. *Astrophys. J.* **481**, 838 (1997)

Hansen, B. M. S. & Phinney, E. S., *Mon. Not. Roy. Astron. Soc.* **291**, 569 (1997)

Heitsch, F., and Richtler, T., *Astron. & Astrophys.* **347**, 455 (1999)

Hesse, J.E., Harris, W.E., Vandenberg, D.A., et al. *Pub. Astron. Soc. Pac.* **739** (1987)

Humphreys, R. M., *Astrophys. J. Suppl.* **38**, 309 (1978)

Johnston, S., Lyne, A. G., Manchester, R. N. , Kniffen, D. A., D'Amico, N., Lim, J. and Ashworth, M., *Mon. Not. Roy. Astron. Soc.* **255**, 401 (1992)

Johnston,S., Manchester, R.N., Lyne, A.G., et al., *Mon. Not. Roy. Astron. Soc.* **268**, 430 (1994)

Kaspi, V.M., taylor, J.H., and Ryba, M.F. *Astrophys. J.* **428**, 713 (1994)

Kaspi, V.M., and Helfand, D.J., In P.O. Slane and B.M. Gaensler, eds., *ASP COnf. Proc. 271: Neutron Stars in Supernova Remnants*, astro-ph/0201183 (2002)

Legge, D, In ASP Conf. Ser. **202**: *IAU Colloq. 177: Pulsar Astronomy - 2000 and Beyond*, **141** (2000)

Longair, M.S., *High Energy Astrophysics*, Second Edition, Cambridge University Press (1994)

Lorimer, D. R., Bailes, M., Dewey, R. J., and Harrison, P. H., *Mon. Not. Roy. Astron. Soc.* **263**, 403 (1993)

Lyne, A. G., Graham- Smith, *Pulsar Astronomy*, Second Edition, Cambridge University Press (1998)

Lyne, A. G., Manchester, R. N., Lorimer, D. R. et al., *Mon. Not. Roy. Astron. Soc.* **295**, 743 (1998)

Lynga, G. *Catalogue of Open Clusters*, 5th ed. (1987)

Manchester, R. N. and Taylor, J. H., *Astron. J.* **86**, 1953 (1981)

Manchester, R. N., Lyne, A. G., D'Amico, N. et al., *Mon. Not. Roy. Astron. Soc.* **279**, 1235 (1996)

Manchester, R. N., Bell, J. F., Camilo, F. et al., *Asp. Conf. Ser.* **271**, 31 (2002)

Morris, D. J., Hobbs, G., Lyne, A. G. et al. *Mon. Not. Roy. Astron. Soc.* **335**, 275 (2002)

Neckel, Th. & Klare, G., *Astron. & Astrophys. Suppl.* **42**, 251 (1980)

Nomoto, K., *Astron. & Astrophys.* **277**, 791 (1984)

Oegelman, H., Koch-Miramond, L., Auriere, M., *Astrophys. J.* **342**, L83 (1989)

Ortolani, S., Barbuy, B., and Bica, E., *Astron. & Astrophys. Suppl.* **108**, 653 (1994)

Ortolani, S., Barbuy, B., and Bica, E., *Astron. & Astrophys.* **308**, 733 (1996)

Paladini, R, Davies, R.D., DeZotti, G., *Mon. Not. Roy. Astron. Soc.* **347**, 237 (2004)

Paltrinieri, B., Ferraro, F.R., Carretta, E., and Fusi Pecci, F., *Mon. Not. Roy. Astron. Soc.* **293**, 434 (1998)

Parker, J. W. & Garmany, C. D., *Astron. J.* **106**, 1471 (1993)

Patel, K. and Pudritz, R. E., *Astrophys. J.* **424**, 668 (1994)

Rees, R.F., and Cudworth, K.M., *Astron. J.* **102**, 152 (1991)

Rey, S., Lee, Y., Byun, Y., and Chun, M., *Astron. J.* **116**, 1775 (1998)

Salpeter, E. E., *Astrophys. J.* **121**, 161 (1955)

Salter, M.J., Lyne, A.G., and Anderson, B., *Nature* **280**, 477 (1979)

Sandquist, E.L., Bolte, M., Stetson, P.B., and Hesser, J.E., *Astrophys. J.* **470**, 910 (1996)

Sarajedini, A., and Norris, J.E., *Astrophys. J. Suppl.* **93**, 161 (1994)

Taylor, J. M., Cordes, J. M. (TC93), *Astrophys. J.* **411**, 674 (1993)

Taylor, J. H., Manchester, R. H., Lyne, A. G., Camilo, F., *Astrophys. J. Suppl.* **88**, 529 (1996)

Trimble, V., and Woltjer, L. *Astrophys. J. Lett.* **163**, 97 (1971)

Weidemann, V., *Ann. Rev. Astron. Astrophys.* **28**, 103 (1990)

Weisberg, J.M., Rankin, J., and Boriakoff, V. *Astron. & Astrophys.* **88**, 84 (1980)

Zombeck, M. V. *Handbook of Space Astronomy and Astrophysics*, Cambridge University Press (1990)

In: Pulsars: Discoveries, Functions and Formation
Editor: Peter A. Travelle, pp. 61-70
ISBN: 978-1-61122-982-0
© 2011 Nova Science Publishers, Inc.

Chapter 4

A LINK BETWEEN THE TYPICAL RADIO PULSARS AND MAGNETARS: MAGNETIC FIELD EVOLUTION THROUGH PULSAR GLITCHES

J.R. Lin[1,*] and S.N. Zhang[1,2,3,4]

[1]Physics Department and Center for Astrophysics,
Tsinghua University, Beijing, China
[2]Physics Department, University of Alabama in Huntsville,
Huntsville, AL,, USA
[3]Space Science Laboratory, NASA Marshall Space Flight
Center, SD50, Huntsville, AL, USA
[4]Institute of High Energy Physics,
Chinese Academy of Sciences, Beijing, China

Abstract

Glitches are thought to be a kind of common phenomena in pulsars these days. From the observations, we know that after each glitch, there is often a permanent increase in the pulsar's spin-down rate (period derivative). Therefore, a pulsar's present spin-down rate may be much higher than its initial value and the estimated characteristic age of a pulsar based on its present spin-down rate and period may be shorter than its true age. At the same time, the permanent increase of its spin-down rate implies that the pulsar's surface magnetic field is increased after each glitch. Consequently, after many glitches, some radio pulsars may evolve into magnetars, i.e., strongly magnetized and slowly rotating neutron stars.

1. Introduction

Pulsars are now accepted to be rapidly rotating and highly magnetized neutron stars formed in supernova explosions. The neutron star emits a steady radio signal that it sweeps around in circles like light from a lighthouse. When the pulsar "lighthouse" is pointing at Earth, we can detect the signal, which would show up as the short pulse. The large mass and

---

*E-mail address: zhangsn@tsinghua.edu.cn

small radius of a neutron star then accounts for the extraordinary stability of the periodical pulses. Pulsars are also characterized by extremely strong magnetic fields. The surface magnetic field of a neutron star may be estimated from the observed period and period derivative, i.e., $B \approx 3.3 \times 10^{19} \sqrt{P\dot{P}}$ (G), if we assume that the pulsar's spin-down energy is overwhelmingly consumed by magnetic dipole radiation, the so-called magnetic-braking model [Pacini 1968].

By the same method, we can also find a distinctive group of pulsars with very high magnetic fields (10^{14}- 10^{15} G) in the pulsars' P - B relation diagram. These pulsars are called magnetars because of their extremely high magnetic fields. Observationally, these pulsars may appear as Soft Gamma-ray Repeaters (SGRs) or Anomalous X-ray Pulsars (AXPs), in which the steady X-ray luminosity is powered by consuming their magnetic field decay energy [Duncan & Thompson 1992, Paczynski 1992, Kouveliotou et al. 1998, Thompson & Duncan 1996, Hurley 2000]. The magnetic field decay also heats the neutron star surface that emits thermal radiation in the X-ray band [Thompson & Duncan 1996]. The high magnetic fields also result in the high spin-down rates, which drive the magnetars to the long period region in rather short time scales. Thus AXPs and SGRs are a small class of pulsars with long periods (5 - 12 s), high spin-down rates and soft X-ray spectra (see Mereghetti 1999 for a comprehensive review). The AXPs and SGRs have young spin-down ages of 10^3 - 10^5 years. Some of them are claimed to be associated with some young supernova remnants (typically 10^3 - 10^4 years old), showing that they may be young pulsars; however the association of these cases are still controversial [Gaensler et al. 2001, Duncan 2002].

Another model for AXPs and SGRs are based on the interaction between the neutron stars and the fossil disks. After the supernova explosion, an amount of mass may fall back onto the central young neutron star [Lin et al. 1991, Chevalier 1989]. In some cases that the materials can not be directly accreted on the surface of the neutron star. When the neutron star's surface magnetic field is strong enough and it also rotates fast enough, the so called magnetosphere radius of the pulsar exceeds the Keplerian co-rotation radius, "propeller" effect takes place. Then the interaction between the fall back material and the magnetosphere of the neutron star somehow causes the neutron star to lose angular momentum and result in timing signatures similar to what is observed for AXPs [Chatterjee et al. 2000, Alpar 2001, Marsden et al. 2001].

In the standard model, pulsars are born with significantly different parameters, and the magnetic fields of typical radio pulsars remain constant or decay slowly during their lifetimes. Thus the radio pulsars will evolve into an "island" which gathers most of the old pulsars. For the magnetars, very high initial dipole magnetic fields are required to slow-down the neutron star to the presently long periods within a relatively short time (10^4 years) when they still have very high magnetic fields. Therefore both the beginning and the ending of typical radio pulsars and the magnetars are very different.

However if we simply assume that all pulsars were born with similar initial parameters and their surface magnetic fields did not change or decay during their subsequent spin-down lives, all observed pulsars should show an overall anti-correlation between their period and spin-down rate. However, in Fig.1, the pulsars (including AXPs and SGRs) with longer spin periods tend to have higher surface magnetic field, as inferred from $B \approx 3.3 \times 10^{19} \sqrt{P\dot{P}}$. Therefore we are forced to the conclusion that either all pulsars were born significantly differently (standard model), or they were born similarly but their surface magnetic fields have been increased during their spin-down lives, resulting in a positive correlation between their surface magnetic fields and spin periods. In this paper, we mainly discuss the latter possibility.

Despite of the distinctive features, AXPs and SGRs shows many similarities with typical radio pulsars, including the properties of glitches. Pulsar glitches (sudden frequency jumps of a magnitude $\frac{\Delta \Omega}{\Omega} \sim 10^{-9}$ to 10^{-6}, accompanied by the jumps of the spin-down rate with a of magnitude $\frac{\Delta \dot{\Omega}}{\dot{\Omega}} \sim 10^{-3}$ to 10^{-2}) are a common phenomenon, which could be well explained by the understanding on the structure of neutron stars and condense matter physics. A relaxation usually happens after a glitch, which tend to reset the pulsar to its status before the glitch. However neither the period nor the spin-down rate rate can be completely recovered. For example, both the Crab pulsar and the Vela pulsar were both found to have a slow increase in their spin-down rates and thus magnetic field increase during the last tens of years [Smith 1999]. Observations show that the glitches happened in the AXPs might cause the huge permanent changes to their spin-down rates [Dall'Osso et al. 2003, Kaspi et al. 2003]. This "unhealed change" in the spin-down rate might be due to the expelled magnetic field from the core to the surface after each glitch, increasing the surface magnetic field of a pulsar [Ruderman et al. 1998]. The model of Ruderman et al. (1998) predicts a certain relation between the glitch rate and the spin-down age for a pulsar. Therefore with the observed glitch parameters and the glitch rate, the long term evolutions of pulsars caused by pulsars' glitches can be calculated. The similarities between AXPs, SGRs and typical radio pulsars make it reasonable to consider the possibility for some typical radio pulsars to evolve to the magnetars, while most other radio pulsars evolve to the "island". In light of the discovery of some normal radio pulsars with long periods and high magnetic fields (comparable to the magnetars) in the Parkes multibeam pulsar survey [Hobbs et al. 2004], it is natural to consider the intrinsic connections between the magnetars and the radio pulsars (especially those with high magnetic field) as an alternative to the previously proposed model [Zhang & Harding 2000].

2. Long Term Evolution of Pulsars Caused by Glitches

In the magnetic-braking model, assuming that the initial period of a pulsar is much smaller than its present value, a pulsar's age may be estimated by its characteristic spin-down age $T_s = \frac{P}{(n-1)\dot{P}}$, where n is the braking index and for the dipole radiation $n = 3$. However for some pulsars, their characteristic ages are much shorter than the ages of the associated supernova remnants. A famous example is PSR B1757-24, a typical radio pulsar with $P = 0.125$ s and $\dot{P} = 1.28 \times 10^{-13}$ s s^{-1}, its characteristic age is 16000 yrs for the braking index $n = 3$. However the associated supernova remnant SNR G5.4-1.2 is believed to be produced between 39000 to 170000 yrs ago [Frail & Kulkarni 1991, Gaensler & Frail 2000, Manchester et al. 2002]. This pulsar was reported to have a very small proper motion in this sky [Thorsett et al. 2002], indicating that either this pulsar should be very old and its magnetic field increased in its history [Thorsett et al. 2002], or this association is wrong. Recently there are some other models supporting the PSR B1757-24 and G5.4-1.2 association [Gvaramadze 2004].

Therefore discrepancy between the spin-down age and associated supernova age for PSR B1757-24 appears to be real. This large discrepancy can not be explained by simply involving a smaller braking index, which in this case would require n<1.2, in contrast to the smallest braking index of n=1.4 (for the Vela pulsar) ever known for all pulsars. However, implied from the measured Ω, $\dot{\Omega}$ and $\ddot{\Omega}$ of this pulsar, the braking index for PSR B1757-24 is $3 - 30$ [Lyne *et al.* 1996]. A fall-back disk model was proposed to explain the age discrepancy [Marsden *et al.* 2001, Shi & Xu 2003]. However a pulsar in a propeller phase should produce a dim thermal x-ray emission, contrary to the observed bright non-thermal emission which is consistent with the standard magnetospheric emission model [Kaspi *et al.* 2001]. Alternatively, the pulsar glitches may be the source of the above age discrepancy. Usually after a glitch, both the period and the spin-down rate of the pulsar are changed, though the period change is usually negligible. However the accumulated increase in the spin-down rate after many glitches will cause a underestimation to the pulsar's age.

In Table.1 we list all the pulsars with both glitches and Supernova Remnants (SNR) reported. All the pulsars and SNRs associations are from the ATNF pulsar catalogue. According to this catalogue, the question marks mean that there are some arguments against the association between the pulsars and the SNRs. We can see that for many glitch pulsars, their true ages are higher than their estimated characteristic ages except those cases with questionable associations. For 3 pulsars with measured permanent change to their spin down rate after glitches, the age discrepancies show roughly positive correlation with the magnitude of the permanent increase of the spin down rate, which we believe could not be a coincidence.

Table 1. Among all glitching radio pulsars which are associated or may be associated with SNRs, most of their characteristic ages are less than the ages of their SNRs, except for two pulsars whose association with SNRs are questionable. The question marks mean that there are some arguments against the association between the pulsars and the SNRs. The ages of the above SNRs are from (a) the record of ancient Chinese observations, (b) [Aschenbach *et al.* 1995], (c) [Caswell *et al.* 1992], (d) [Dodson & Golap 2002], (e) [Gaensler & Frail 2000], (f) [Rho & Borkowski 2002], (g) [Finley & Ogelman 1994], (h) [Blanton & Helfand 1996]. The average value of $\frac{\Delta\dot{\nu}_p}{\dot{\nu}_p}$ for three pulsars were implied from [Lyne *et al.* 1993], [Wang *et al.* 2001], [Wong *et al.* 2001], [Lyne *et al.* 2000], [Dodson *et al.* 2002] and [Lyne *et al.* 1996]

Pulsar name	Characteristic age (kyr)	associated SNR	SNR age (kyr)	$\frac{\Delta\dot{\nu}_p}{\dot{\nu}_p}$
B0531+21	1.258	Crab	0.949 (a)	0.00009
B0833-45	11	Vela	18 (b)	0.00092
B1338-62	12	G308.8-0.1	32.5 (c)	-
B1706-44	17	G343.1-2.3(?)	8.9(d)	-
B1757-24	16	G5.4-1.2	39 - 170 (e)	0.0037
B1758-23	58	W28(?)	36 (f)	-
B1800-21	16	G8.7-0.1(?)	10 - 50 (g)	-
J1846-0258	0.72	Kes75	0.9 - 4.3 (h)	-

The possibility for the magnetic growth [Blandford *et al.* 1983] and to reconcile the age discrepancy of some pulsars by the magnetic field growth [Chanmugam *et al.* 1995] or pulsars' glitches [Marshall *et al.* 2004] have been discussed previously. With the data of the observed glitch parameters of some pulsars, we can do quantitatively more detailed calculations. In this work we investigate the roles of pulsar's glitches in the long term evolution of pulsars, in order to explain the observed positive correlation between pulsars' surface magnetic fields and their periods, and the large discrepancy between pulsars' characteristic ages and their associated supernova remnants.

We take PSR B1757-24 as an example to illustrate the long term evolutionary effects caused by pulsars' glitches. A giant glitch in PSR B1757-24 was reported [Lyne *et al.* 1996]. The long-term post-glitch relaxation fit shows that $\frac{\Delta \dot{P}_p}{\dot{P}} \approx 0.0037$, where $\Delta \dot{P}_p$ is the permanent change to $\dot{P}$ after the glitch. Based on the observational fact that different pulsars have on the average distinctive $\frac{\Delta \dot{P}_p}{\dot{P}}$, it is not unreasonable to assume that similar giant glitches have happened repeatedly in the history of PSR B1757-24, therefore we can describe its spin-down history by the following three equations:

$$P - P_0 = \int_0^\tau \dot{P}(t)\mathbf{d}t = \sum \int_0^{\tau_n} \dot{P}_n(t)\mathbf{d}t. \tag{1}$$

where τ_n is the interval between two adjacent glitches. The previous observations indicate that the glitch activity may be negatively related to the pulsar's spin-down age except for the youngest pulsars such as the Crab pulsar [Lyne *et al.* 1995]. Both this relation and the exception can be well explained by a theoretical work in which $\tau_n \propto \frac{P}{\dot{P}} \propto T_s$ for a pulsar [Ruderman *et al.* 1998]. We adopt this relation in our calculations. Assuming that between two adjacent glitches, the surface magnetic field of the pulsar remains constant, we have

$$\dot{P}_n(t)P_n(t) = \mathbf{Const}_n \tag{2}$$

For PSR B1757-24 we assume the following relationship is true for every glitch:

$$\dot{P}_{n+1}(\tau_n) = \alpha \dot{P}_n(0), \tag{3}$$

where $\alpha = 1.0037$ for PSR B1757-24.

Taking its present values of $P = 0.125$ s and $\dot{P}_n = 1.28 \times 10^{-13}$ s s^{-1} and assuming that its true age is $\tau = 10^5$ yrs and its initial period is $P_0 = 10$ ms, we get $n = 1495$ from the above equations and its initial surface magnetic field is about 2.6×10^{11} Gauss. We can also estimate that its glitch rate would be about once per 3.4 years when it was 1000 years old, similar with the observed glitch rate for the Crab pulsar. If we assume that future glitch rate for PSR B1757-24 will continue to follow this pattern, then after 2×10^5 years, its characteristics will be similar to AXPs as shown in Fig.1. Assuming that all pulsars were born with the same surface magnetic field and spin period, but different glitch properties, their different evolutionary paths are also shown in Fig.1 for different values of $\frac{\Delta \dot{P}_p}{\dot{P}}$. The equal-age lines are also shown in the figure, in contrast to the characteristic ages of pulsars based simply on their present day period and spin-down rate without taking into account of pulsar glitches. Under the same assumption, in Fig.2 we show the estimated ages for the pulsars with given P and $\dot{P}$.

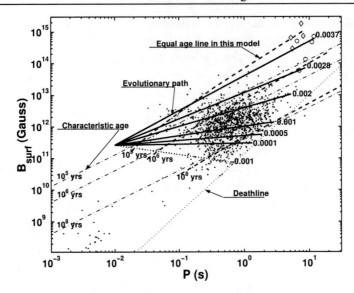

Figure 1. All known isolated pulsars are shown in this figure. Filled circles are millisecond and "regular" radio pulsars, open circles are AXPs, and diamonds are SGRs. The pulsars with longer spin periods tend to have higher surface magnetic field, as inferred from $B \approx 3.3 \times 10^{19}\sqrt{P\dot{P}}$. The dotted-line is the "death-line" for radio pulsars [Chen & Ruderman 1993]. The magnetic field evolution of pulsars are caused by the permanent changes to the spin-down rates after glitches. Different solid lines denote different values of $\frac{\Delta\dot{P}_p}{\dot{P}}$. The line for $\frac{\Delta\dot{P}_p}{\dot{P}} = 0.0037$ is the evolutionary path of PSR B1757-24, calculated from equations (1), (2) and (3). Assuming that all pulsars were born with the same initial surface magnetic field and spin period, but different glitch properties, their different evolutionary paths are also shown for different values of $\frac{\Delta\dot{P}_p}{\dot{P}}$. Pulsars on the same dashed-lines have the same age as calculated in our model, in contrast to the characteristic ages (dashed-dotted lines) of pulsars based simply on their present day period and spin-down rate without taking into account of pulsar glitches.

3.   Conclusions

Based on our results presented above, we have the following conclusions and remarks:

(1) Pulsar glitches, especially the permanent changes to their spin-down rates, are important for the long term evolutions of pulsars. The previous evolutionary history and future "fate" of a pulsar may be calculated within the frame work of the magnetic-braking model, if its glitch properties are known.

(2) If we assume all pulsars were born similarly, then the positive $P - B$ correlation may be explained naturally. However any slight differences in the initial conditions for different pulsars may cause a large uncertainties for their age estimates when pulsars are younger than 10^5 years , as seen in Fig.1 and Fig.2; the age estimated from our model is reliable only for those pulsars older than 10^5 years, such as PSR B1757-24.

(3) Our model suggests that some radio pulsars, such as PSR B1757-24 and other pulsars which exhibit large values of $\frac{\Delta\dot{P}_p}{\dot{P}}$ and are thus along similar evolutionary paths

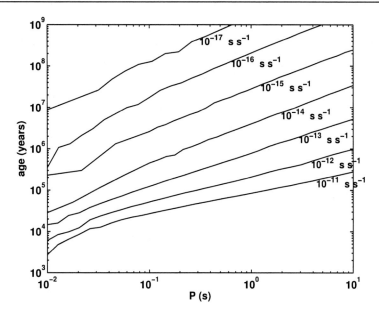

Figure 2. Assuming the same initial condition, for given P and $\dot{P}$ we can calculate the pulsar's age from equations (1), (2) and (3). Since the effect caused by the glitches is a accumulated process, our model for pulsar's age estimate is reliable only for pulsars older than 10^5 years, but has considerable uncertainties for younger pulsars. The solid lines are for the different values of $\dot{P}$.

shown in Fig. 1, may eventually evolve into AXPs and SGRs within 10^5 to 10^6 years after their birth, contrary to the classical "magnetar" model in which they are very young neutron stars born with very high magnetic fields [Duncan & Thompson 1992]. Our model requires that AXPs and SGRs have glitches with large values of $\frac{\Delta \dot{P}_p}{\dot{P}}$ and a glitch rate of once per several years; this is consistent with the observed glitch properties of AXPs [Dall'Osso *et al.* 2003, Kaspi *et al.* 2003]. However, if the associations of young SNRs with some magnetars are true (however see [Gaensler *et al.* 2001] for arguments against the associations), we can not rule out the possibility that some magnetars might be born with ultra-high magnetic field, or their glitch histories are significantly different from known radio pulsars.

(4) In our model, the small number of known magnetars compared to "regular" radio pulsars requires their progenitors should also be rare. This is roughly consistent with the small number of pulsars along the evolutionary paths leading to the magnetars, as can be seen in Fig.1. Since the values of $\frac{\Delta \dot{P}_p}{\dot{P}}$ decide the pulsars' fates, we can estimate the expected percentage of magnetars with Fig.1. In Fig.1, only the pulsars with $\frac{\Delta \dot{P}_p}{\dot{P}} > 0.0028$ would evolve to the magnetars. So among all these isolated pulsars in this diagram, 4.6% of them will evolve into magnetars. In the P-B diagram, there is an observational region for the AXPs and the SGRs. Estimation based on our model shows that the time scale for the pulsars to go through this region is 1.5×10^4 yrs and the total lifetime (from the initial point we set in our model to the end of the magnetar phase) for PSR B1757-24 is 2×10^5 yrs. Therefore, under the assumption of a uniform pulsar birth rate over time, for

the pulsars that will evolve into magnetars, around 7.5% of their lives will be in the magnetar phase. Currently it is difficult to compare accurately the expected number of magnetars with known magnetars, because the samples for both radio pulsars along the evolutionary path and magnetars may be quite incomplete.

(5) Our model does not include the long term magnetic field decay (MFD) of pulsars [Gunn & Ostriker 1970]. However for pulsars with active glitches, the magnetic field increase by glitches overwhelms the slow magnetic field decay. In the case that the significant magnetic field decay is inevitable such as the magnetars whose X-ray emission is believed to be powered by the magnetic field decay energy, our model infers a time scale of only 3×10^4 years for the magnetic field to increase from 10^{14} Gauss to 10^{15} Gauss. In contrast, the estimated time scale of MFD (induced by the Hall cascade) from 10^{14} Gauss to 10^{13} Gauss is 10^5 years, and it takes more than 10^7 years for the same amount of decay driven by ambipolar diffusion [Colpi *et al.* 2000]. A more realistic model for pulsars with "weak" glitch properties should also include the long term magnetic field decay process. We will investigate this in the future.

(6) In Fig.2 the ages are estimated for most pulsars according to their period and period derivative by equations (1), (2) and (3). These predictions may be tested with future pulsar and SNR observations.

(7) Finally we should mention that since our model does not assume different radiation mechanisms for all pulsars, the birth and death lines for pulsars remain unchanged.

Acknowledgment

We thank Drs. Zigao Dai, Yang Chen, Renxin Xu and Guojun Qiao for valuable comments to the manuscript. This study is supported in part by the Special Funds for Major State Basic Research Projects (10233010) and by the National Natural Science Foundation of China. SNZ also acknowledges supports by NASA's Marshall Space Flight Center and through NASA's Long Term Space Astrophysics Program.

References

[Aschenbach *et al.* 1995] Aschenbach, B., Egger, R. & Trumper, J. *Nature*, **373**, 587 1995

[Alpar 2001] Alpar, M.A. *Astrophys. J.*, **554**, 1245 2001

[Blandford *et al.* 1983] Blandford, R. D., Applegate, J. H., & Hernquist, L. *MNRAS*, **204**, 1025 1983

[Blanton & Helfand 1996] Blanton, E. L. & Helfand, D. J. *Astrophys. J.*, **470**, 961 1996

[Caswell *et al.* 1992] Caswell, J. L., Kesteven, M. J., Stewart, R. T., Milne, D. K. & Haynes, R. F. *Astrophys. J.*, **399**, L151 1992

[Chanmugam *et al.* 1995] Chanmugam, G., Rajasekhar, A. & Young, E. J. *MNRAS*, **276**, L21 1995

[Chatterjee *et al.* 2000] Chatterjee, P., Hernquist, L., & Narayan, R. *Astrophys. J.*, **534**, 373 2000

[Chen & Ruderman 1993] Chen, K. & Ruderman, M. *Astrophys. J.*, **402**, 264 1993

[Chevalier 1989] Chevalier, R.A., *Astrophys. J.*, **346**, 847 1989

[Colpi *et al.* 2000] Colpi, M., Geppert, U. & Page, D. *Astrophys. J.*, **529**, L29 2000

[Dall'Osso *et al.* 2003] Dall'Osso, S., Israel, G. L., Stella, L., Possenti, A. & Perozzi, E. *Astrophys. J.*, **599**, 485 2003

[Dodson & Golap 2002] Dodson, R. G. & Golap, K. *MNRAS*, **334**, L1 2002

[Dodson *et al.* 2002] Dodson, R. G., McCulloch, P. M. & Lewis, D. R. *Astrophys. J.*, **564**, L85 2002

[Duncan & Thompson 1992] Duncan, R. C. & Thompson, C. *Astrophys. J.*, **392**, L9 1992

[Duncan 2002] Duncan, R. C. *Mem. Soc. Astron. Italiana*, **73**, 534 2002

[Finley & Ogelman 1994] Finley, J. P. & Ogelman, H. *Astrophys. J.*, **434**, L25 1994

[Frail & Kulkarni 1991] Frail, D. A, & Kulkarni, S. R. *Nature*, **352**, 785 1991

[Gaensler & Frail 2000] Gaensler, B. M. & Frail, D. A. *Nature*, **406**, 158 2000

[Gaensler *et al.* 2001] Gaensler, B. M., Slane, P. O., Gotthelf, E. V. & Vasisht, G. *Astrophys. J.*, **559**, 963 2001

[Gunn & Ostriker 1970] Gunn, J. E., & Ostriker, J. P. *Astrophys. J.*, **160**, 979 1970

[Gvaramadze 2004] Gvaramadze, V. V. *A&A*, **415**, 1073 2004

[Hobbs *et al.* 2004] Hobbs, G., Faulkner, A., Stairs, I. H., Camilo, F., Manchester, R. N., Lyne, A. G., Kramer, M., D'Amico, N., Kaspi, V. M., Possenti, A., McLaughlin, M. A., Lorimer, D. R., Burgay, M., Joshi, B. C. & Crawford, F., *MNRAS*, **352**, 1439 (2004)

[Hurley 2000] Hurley, K., in AIP Conf. Proc. 526, Fifth Huntsville Symp. on Gamma-Ray Bursts, ed. R. M. Kippen, R. S. Mallozzi, & G. J. Fishman (New York:AIP), 763

[Kaspi *et al.* 2001] Kaspi, V. M., Gotthelf, E. V., Gaensler, B. M. & Lyutikov, M. *Astrophys. J.*, **562**, L63 2001

[Kaspi *et al.* 2003] Kaspi, V. M. Gavriil, F. P., Woods, P. M., Jensen, J. B., Roberts, M. S. E., & Chakrabarty, D. *Astrophys. J.*, **588**, L93 2003

[Kouveliotou *et al.* 1998] Kouveliotou, C., Dieters, S., Strohmayer, T., van Paradijs, J., Fishman, G. J., Meegan, C. A., Hurley, K., Kommers, J., Smith, I., Frail, D., & Murakami, T. *Nature*, **393**, 235 1998

[Lin *et al.* 1991] Lin, D.N.C., Woosley, S.E. & Boddenheimer, P.H. *Nature*, **353**, 827 1991

[Lyne *et al.* 1993] Lyne, A. G., Pritchard, R. S. & Smith, F. G. *MNRAS*, **265**, 1003 1993

[Lyne *et al.* 1995] Lyne, A. G., Pritchard, R. S. & Shemar, S. L. *J. Astrophys. Astron.*, **16**, 179 1995

[Lyne *et al.* 1996] Lyne, A. G., Kaspi, V. M., Bailes, M., Manchester, R. N., Taylor, H. & Arzoumanian, Z. *MNRAS*, **281**, L14 1996

[Lyne *et al.* 2000] Lyne, A. G., Shemar, S. L. & Smith, F. G. *MNRAS*, **315**, 534 2000

[Manchester *et al.* 2002] Manchester, R. N., Bell, J. F., Camilo, F., Kramer, M., Lyne, A. G., Hobbs, G. B., Joshi, B. C., Crawford, F., D'Amico, N., Possenti, A., Kaspi, V. M. & Stairs, I. H., *Young Pulsars from the Parkes Multibeam Pulsar Survey and their Associations. Neutron Stars in Supernova Remnants*, **31**, eds Slane, P. O. & Gaensler, B. M., Astronomical Society of the Pacific, San Francisco 2002

[Marsden *et al.* 2001] Marsden, D., Lingenfelter, R. E. & Rothschild, R. E. *Astrophys. J.*, **547**, L45 2001

[Marshall *et al.* 2004] Marshall, F. E., Gotthelf, E. V., Middleditch, J., Wang, Q. D. & Zhang, W. *Astrophys. J.*, **603**, 682 2004

[Mereghetti 1999] Mereghetti, S. invited review presented at NATO ASI, The Neutron Star-Black Hole Connection (Elounda, Crete) (astro-ph/9911252) 1999

[Pacini 1968] Pacini, F. *Nature*, **219**, 145 1968

[Paczynski 1992] Paczynski, B., *Acta Astron.*, **42**, 145

[Rho & Borkowski 2002] Rho, J. & Borkowski, K. J. *Astrophys. J.*, **575**, 201 2002

[Ruderman *et al.* 1998] Ruderman, M., Zhu, T. & Chen, K. *Astrophys. J.*, **492**, 267 1998

[Shi & Xu 2003] Shi, Y., & Xu, R. X, *Astrophys. J.*, **596**, L75 2003

[Smith 1999] Smith, F. G. *in Pulsar Timing, General Relativity and the Internal Structure of Neutron Stars*, ed. Z. Arzoumanian, F. Van der Hooft, & E. P. J. van den Heuvel (Amsterdam: Koninklijke Nederlandse Akademie van Wetenschappen), P. 151. 1999

[Thompson & Duncan 1996] Thompson C. & Duncan R.C. *Astrophys. J.*, **473**, 322, 1996

[Thorsett *et al.* 2002] Thorsett, S. E., Brisken, W. F. & Goss, W. M. *Astrophys. J.*, **573**, L111 2002

[Wang *et al.* 2001] Wang, N., Wu, X., Manchester, R. N., Zhang, J., Lyne, A. G. & Yusup, A. *Chin. J. Astron. Astrophys*, Vol. 1, No. 3, 195 2001

[Wong *et al.* 2001] Wong, T., Backer, D. C. & Lyne, A. G. *Astrophys. J.*, **548**, 447 2001

[Zhang & Harding 2000] Zhang, B. & Harding, A. K. *Astrophys. J.*, **535**, L51 2000

In: Pulsars: Discoveries, Functions and Formation
Editor: Peter A. Travelle, pp. 71-95

ISBN 978-1-61122-982-0
© 2011 Nova Science Publishers, Inc.

Chapter 5

NEUTRINOSPHERES, RESONANT NEUTRINO OSCILLATIONS, AND PULSAR KICKS

M. Barkovich[1], J.C. D'Olivo[1] and R. Montemayor[2]
[1] Departamento de Física de Altas Energías,
Instituto de Ciencias Nucleares, Universidad Nacional
Autónoma de México, Apartado Postal 70-543,
04510 México, Distrito Federal, México.
[2] Instituto Balseiro and Centro Atómico Bariloche,
Universidad Nacional de Cuyo and CNEA,
8400 S. C. de Bariloche, Río Negro, Argentina.

Abstract

Pulsars are rapidly rotating neutron stars and are the outcome of the collapse of the core of a massive star with a mass of the order of or larger than eight solar masses. This process releases a huge gravitational energy of about 10^{53} erg, mainly in the form of neutrinos. During the collapse the density increases, and so does the magnetic field due to the trapping of the flux lines of the progenitor star by the high conductivity plasma. When the density reaches a value of around $10^{12}\,\mathrm{g\,cm^{-3}}$ neutrinos become trapped within the protoneutron star and a neutrinosphere, characterized inside by a diffusive transport of neutrinos and outside by a free streaming of neutrinos, is formed and lasts for a few seconds. Here we focus on the structure of the neutrinosphere, the resonant flavor conversion that can happen in its interior, and the neutrino flux anisotropies induced by this phenomena in the presence of a strong magnetic field. We present a detailed discussion in the context of the spherical Eddington model, which provides a simple but reasonable description of a static neutrino atmosphere, locally homogenous and isotropic. Energy and momentum are transported by neutrinos and antineutrinos flowing through an ideal gas of nonrelativistic, nondegenerate nucleons and relativistic, degenerate electrons and positrons. We examine the details of the asymmetric neutrino emission driven by active-sterile neutrino oscillations in the magnetized protoneutron star, and the possibility for this mechanism to explain the intrinsic large velocities of pulsars respect to nearby stars and associated supernova remnants.

1. Introduction

The finding of pulsars in the sixties was one of the most astonishing discoveries in astrophysics[1]. Soon after their observation they were identified with neutron stars[2], whose existence had been theoretically predicted decades before. Neutron stars are the densest, most rapidly rotating, and most strongly magnetized objects in the Galaxy, features well understood by the standard gravitational and elementary particle physics[3]. They are unique laboratories, which give us access to extreme conditions that are virtually impossible to obtain on Earth. Neutron stars are of particular interest for neutrino physics because of the key role that these particles play in the very first stages of their formation[4].

There is a remarkable characteristic of pulsars not satisfactorily understood: their large drift velocities with respect to nearby stars. The average value of pulsar velocities is in the range of $200 - 500 \, \mathrm{km \, s^{-1}}$, an order of magnitude larger than the mean velocity of ordinary stars in our Galaxy ($\sim 30 \, \mathrm{km \, s^{-1}}$)[5]. In addition, the distribution of these velocities is not Gaussian. It seems to have a bimodal structure[6], with a significant fraction ($\sim 15\%$) of the pulsar population having velocities higher than $1000 \, \mathrm{km \, s^{-1}}$ and up to a maximum of $1600 \, \mathrm{km \, s^{-1}}$.

Neutron stars are formed in type-II supernova explosions and, given the enormous energy liberated during these processes, it is quite natural to look for an explanation in terms of an impulse or *kick* received during their birth. This hypothesis is supported by recent analysis of individual pulsar motions and of associations between supernovae remnants and pulsars. It is also consistent with observed characteristics of binary systems with one or both constituents being a neutron star[7]. Despite this evidence, the physical origin of the kick is one of the most important unsolved puzzles in supernovae research. If the distribution is bimodal, then more than one mechanism could be responsible for the natal kicks. Two classes of mechanism have been considered[8]. One of them invokes an asymmetric mass ejection during the supernova explosion as the result of hydrodynamic perturbations caused by global density inhomogeneities prior to the core collapse[9] or convective instabilities in the neutrino-heated layer behind the shock[10]. Recoil speeds as high as $\sim 500 \, \mathrm{km \, s^{-1}}$ have been obtained with two-dimensional simulations. However, a recent three-dimensional calculation shows that even the most extreme asymmetric collapse does not produce final neutron star velocities above $200 \, \mathrm{km \, s^{-1}}$[11], rendering the situation unclear.

Only a small fraction of the gravitational binding energy ($E \simeq 10^{53} \, \mathrm{erg}$) liberated during a supernova explosion is required to account for the huge electromagnetic luminosity observed and the ejection of the envelope. More than 99% of the energy is released in the form of neutrinos, which are copiously produced when the iron core turns into a neutron core. A small asymmetry ($\sim 1\%$) in the momentum taken away by neutrinos should be enough to give the nascent pulsar an acceleration consistent with the measured velocities. The asymmetric neutrino emission induced by strong magnetic fields is the basis of the second class of kick mechanisms, and effects such as parity violation[12], asymmetric field distributions[13], and dark spots[14] have been considered in this context.

The density in the interior of a protoneutron star is so high, of the order of the nuclear density, that even neutrinos remain trapped, being emitted from an approximately well defined surface, the neutrinosphere. In a magnetized stellar plasma there are anisotropic contributions to the neutrino refraction index[15, 16] and, as a consequence, the resonant flavor transformations of neutrinos diffusing in different directions respect to the magnetic field occur at different depths within the protostar. This phenomenon can induce a momentum flux asymmetry and is the basis of the neutrino-driven kick mechanism proposed by Kusenko and Segrè (KS)[17, 18, 19]. With standard (active) neutrinos the mechanism works when the resonance region lies between the ν_e and ν_μ (or ν_τ) neutrinospheres, but requires an exceedingly high square mass difference, in conflict with the existing limits. A variant avoiding such a limitation can be implemented in terms of active-sterile neutrino oscillation[20, 21, 22].

The measurement of the Z-boson width at LEP, limiting to three the number of light neutrinos having weak interactions, does not preclude the existence of additional neutral fermions that do not fill the standard model gauge interactions. They have been invoked to explain several questions in astronomy and cosmology[23] and may becomes a necessity if the claimed LSND evidence for oscillations were confirmed. The range of the oscillation parameters (mass and mixing angle) required by the pulsar kick mechanism overlaps with the allowed region for sterile neutrinos being candidates to cosmological dark matter. Alternative kick mechanisms driven by neutrino oscillations have been proposed invoking the possible existence of transition magnetic moments[24], non orthonormality of the flavor neutrinos[25], violations of the equivalence principle[26], or off-resonant emission of sterile neutrinos[21].

In this article we review the explanation of the pulsar kick in terms of the resonant neutrino oscillations. In Section II, we outline the creation of a protoneutron star as a result of a supernova event, and the formation in its interior of a gas of trapped neutrinos. The momentum flux carried by neutrinos is analyzed in Section III. Section IV is devoted to the discussion of the neutrino oscillations in matter incorporating the effect of strong magnetic fields. The implication of this phenomenon as source of the asymmetry in the momentum flux transported by neutrinos emitted during the protoneutron star cooling is examined in Section V. Numerical results are obtained in Section VI within the context of the Eddington model, where the kick effect is explicitly evaluated. In the final section we summarize the outstanding features of the mechanism.

2.   Stellar Collapse, Protoneutron Stars, and Neutrinospheres

Neutron stars are extremely compact objects born in type II supernova explosions[27], as the aftermath of the gravitational collapse of a massive star ($M \gtrsim 8 M_\odot$) when it exhausts the sources of thermonuclear energy and its iron core reaches the Chandrasekhar limit, $M_{Ch} = 1.44\,M_\odot$. The process is triggered by the photodissociation of iron nuclei, $\gamma + {}^{56}\text{Fe} \rightarrow 13\,{}^4\text{He} + 4n - 124, 4\,\text{MeV}$, which consumes energy reducing the pressure of the degenerated electron gas. When the collapse proceeds, pressure support decreases because of the electron capture by nuclei. Electron neutrinos are copiously produced through the reaction $p + e \rightarrow n + \nu_e$ and initially they freely escape from the core, but as density increases the medium becomes less transparent for them. The inner core ($M \approx 0.5 - 0.8\,M_\odot$) collapses in a homologous way at a subsonic velocity proportional to the radius ($v/r = 400 - 700\,\text{s}^{-1}$), with the density and temperature profiles remaining similar and only the scale changing with time[28]. The external shells collapse at a supersonic velocity of the order of free fall velocity ($v \propto 1/\sqrt{r}$).

During the first part of the collapse the main source of opacity for neutrinos is the coherent scattering by heavy nuclei, mediated by the neutral current weak interaction. The cross section for this reaction is proportional to A^2, where A is the number of nucleons in the nuclei, and is very effective in trapping neutrinos[29]. If λ_ν is the mean free path of neutrinos, then the time for them to diffuse out of the core, estimated from a random walk through a distance of the order of the core radius R, is $t_{diff} \sim 3R^2/\lambda_\nu$. On the other hand, the core dynamical time scale is of the order of the free fall time $t_c \sim (G\rho)^{-1/2}$. At a certain density, of about 10^{12} g cm^3 for 10 MeV neutrinos, both times become comparable and neutrinos are trapped[30]. β-equilibrium between e and ν_e is set up and electron neutrinos also become degenerated. In this way, the collapsing core is the only place, besides the early universe, where neutrinos are in thermal equilibrium. During the collapse the lepton fraction $Y_L = Y_e + Y_{\nu_e}$ is kept almost constant near the value of the electron fraction at the beginning of the trapping ($Y_e \approx 0.35$), with only $\sim 1/4$ corresponding to neutrinos[4].

The degenerate gas of trapped neutrinos diffuses towards the exterior, reaching regions of decreasing density where the collisions are less frequent, until they finally decouple from the star. Although this process is continuous, it is convenient to introduce an effective emission surface called the neutrinosphere, characterized as the surface where the optical depth becomes approximately one. More accurately, the radius of the neutrinosphere R_{ν_ℓ} is defined by the condition

$$\int_{R_{\nu_\ell}}^{\infty} \frac{dr}{\lambda_{\nu_\ell}} \sim \frac{2}{3}. \tag{1}$$

The medium is considered opaque for neutrinos inside the neutrinosphere and transparent outside. We can also define the energy sphere, where the energy exchanging reactions freeze out while energy conserving collisions may still be important.

The collapse stops suddenly when, after a fraction of seconds, densities of around 3×10^{14} g cm^{-3}, exceeding nuclear density, are reached[31]. Then, a shock wave is built up near the interface of the halted inner core and the supersonically free falling outer part. The shock wave propagates outwards and eventually ejects the stellar mantle of the star. At its passage through the neutrinosphere nuclei are dissociated decreasing the neutrino opacity and causing a recession of the neutrinosphere. The ν_e produced by the reaction $p + e \rightarrow n + \nu_e$ can escape freely building up a short (< 10 msec) ν_e burst, called the "neutronization burst" or "deleptonization burst". The star remnant gravitationally decouples from the expanding ejecta and a protoneutron star has been formed[32].

The neutrinos continue trapped due to the interaction with the residual particles of the medium, mainly nucleons, electrons, and neutrinos. For ν_e the dominant reactions are the β and β-inverse, which mantain them in local thermodynamical equilibrium[33]. In this case, the distinction between neutrinosphere and energy sphere is not crucial. The electron neutrinos also interact with nucleons through the neutral current, but the cross sections for these reactions are smaller. Although the energy dependence of the process involving ν_e and $\bar{\nu}_e$ is the same, in a protoneutron star the number of neutrons is larger than the number of protons, and therefore the ν_e suffer more collisions than the $\bar{\nu}_e$. As a consequence, the neutrinosphere of the ν_e is a bit further out than the one of the $\bar{\nu}_e$, and the mean energy per emitted particle is smaller. After the achievement of β equilibrium, the total entropy is conserved and the collapse becomes adiabatic.

In the case of the nonelectron (active) neutrinos and antineutrinos, the absence of muons and taus does not allow the establishment of chemical equilibrium with the stellar plasma. Therefore, they do not transport leptonic number outside the star and their chemical potentials vanish. Their interactions with the background are the same and no distinction needs to be made between ν_μ and ν_τ. Deeper in the protostar the main interactions for these neutrinos are pair processes such as $\mathcal{N}\mathcal{N}' \rightleftarrows \mathcal{N}\mathcal{N}'\nu_{\mu,\tau}\bar{\nu}_{\mu,\tau}$, $e^-e^+ \rightleftarrows \nu_{\mu,\tau}\bar{\nu}_{\mu,\tau}$, etc, where $\mathcal{N}$ denotes a nucleon, and the scattering reactions with nucleons and electrons. The scattering on electrons are less frequent, but the energy transferred is greater. This process dominates the energy exchange up to the energy sphere where it freezes out. Beyond this radius, muon and tau neutrinos still scatter off neutrons until they reach the neutrinosphere. The cross section for $\mathcal{N}\nu_{\mu,\tau} \rightleftarrows \mathcal{N}\nu_{\mu,\tau}$ is smaller than the one of the β reaction, and hence the radius of the muon (tau) neutrinosphere $R_{\nu_{\mu,\tau}}$ is shorter than the radius of the electron neutrinosphere R_{ν_e}. The protoneutron star contracts, cools, and deleptonizes due to the emission of neutrinos and antineutrinos of all the flavors in a scale of time of the order of 10 sec. The released energy is equally shared by all neutrino and antineutrino species. This phase is known as the Kelvin-Helmholtz phase and ends up in the formation of a neutron star. During this time the radius of the protostar varies from around 30 to 10 km, while the mass remains of the order of $1.4\,M_\odot$. After approximately 1 minute the neutrino mean free path becomes comparable to the stellar radius and neutrino luminosity decreases quickly.

Another important property predicted for neutron stars is that they should have an extremely strong magnetic field. Since the conductivity of the iron core is very large, the magnetic flux is essentially trapped and it is conserved as it collapses to form a neutron star[34]. Adopting for the initial magnetic field the typical values observed in some white dwarfs, from the flux conservation one can easily estimate that the surface field of a protoneutron star has to be in the range of $10^{12} - 10^{13}$ G. This is supported by evidence inferred from measurements of the spin-down rates and observations of the X-ray spectra features[35]. Magnetic fields of the order of 10^{15} G or stronger can be generated by dynamo processes, and in the central regions intensities up to 10^{18} G have been considered[36].

3. Momentum Flux

In a protoneutron star the momentum flux from the core to the external regions is dominated by neutrinos, which provide the most efficient mechanism to release the energy generated during stellar collapse. As explained in the previous section, matter densities are so high that neutrinos are trapped forming a degenerated gas at quasi-equilibrium diffusing radially. The statistical description of this gas is made in terms of a distribution function $f_{\nu_\ell}(\mathbf{x}, \mathbf{k}, t)$. If the action of external fields is negligible, then the neutrino momentum between two consecutive collisions remains constant. The change of the distribution function is due to neutrino collisions with the particles in the medium and is described by the Boltzmann equation

$$\frac{df_{\nu_\ell}}{dt} = C(f_{\nu_\ell})\,, \tag{2}$$

where $C(f_{\nu_\ell})$ is the collision integral and

$$\frac{df_{\nu_\ell}}{dt} = \frac{\partial f_{\nu_\ell}}{\partial t} + \hat{\mathbf{k}} \cdot \nabla f_{\nu_\ell}\,. \tag{3}$$

When the regime is close to equilibrium we can approximate the collision integral as follows

$$C(f_{\nu_\ell}) \simeq -\frac{f_{\nu_\ell} - f_{\nu_\ell}^{eq}}{\tau_{\nu_\ell}}, \tag{4}$$

where $\tau_{\nu_\ell} \cong \lambda_{\nu_\ell}$ is the average time between collisions and

$$f_{\nu_\ell}^{eq}(k) = \frac{1}{1 + e^{(k - \mu_{\nu_\ell})/T}} \tag{5}$$

is the Fermi-Dirac distribution function for the gas of relativistic neutrinos ($E_{\nu_\ell} \cong k = |\mathbf{k}|$) with chemical potential μ_{ν_ℓ} and temperature T. In the static situation ($\partial f_{\nu_\ell}/\partial t = 0$), Eq. (2) with $C(f_{\nu_\ell})$ given by Eq. (4) can be solved iteratively. Proceeding in this way we obtain

$$f_{\nu_\ell} \simeq f_{\nu_\ell}^{eq} - \lambda_{\nu_\ell}\hat{\mathbf{k}} \cdot \nabla f_{\nu_\ell}^{eq}, \tag{6}$$

which constitutes the diffusion approximation for the distribution function. The second term in this expression gives the correction to equilibrium and allows the transport of the energy and momentum towards the exterior of the protostar.

Once the distribution function is known the neutrino energy-momentum tensor can be computed as

$$T_{\nu_\ell}^{\mu\nu} = \int \frac{d^3k}{(2\pi)^3} \frac{k^\mu k^\nu}{k^0} f_{\nu_l}. \tag{7}$$

From this formula, for the energy density we have

$$U_{\nu_l} = T_{\nu_l}^{00} = \frac{1}{(2\pi)^3} \int k^3 dk \sin\theta' d\theta' d\phi' f_{\nu_l} = \frac{F_3[\eta_{\nu_l}]}{2\pi^2} T^4, \tag{8}$$

where $\eta_{\nu_l} = \mu_{\nu_l}/T$ is the degeneracy parameter and $F_3[\eta_{\nu_l}]$ is a Fermi integral

$$F_n[\eta_{\nu_l}] = \int_0^\infty dx\, x^n [1 + \exp(x - \eta_{\nu_l})]^{-1}. \tag{9}$$

At each point within the protostar we choose a local frame (x', y', z') with $\hat{\mathbf{r}} = \hat{\mathbf{z}}'$, such that $\hat{\mathbf{k}} = \cos\theta'\hat{\mathbf{z}}' + \sin\theta'\cos\phi'\hat{\mathbf{x}}' + \sin\theta'\sin\phi'\hat{\mathbf{y}}'$.

The stress tensor can be expressed in terms of the energy density as

$$T_{\nu_l}^{ij} = \delta^{ij}\frac{U_{\nu_l}}{3}. \tag{10}$$

The equilibrium component of the distribution function is isotropic and gives a null contribution to the momentum flux ($F_{\nu_l}^i = T_{\nu_l}^{0i}$), which is generated exclusively by the diffusive term:

$$\mathbf{F}_{\nu_\ell} = \frac{1}{(2\pi)^3} \int k^2 dk \sin\theta' d\theta' d\phi'\, \mathbf{k} f_{\nu_\ell} = -\frac{1}{6\pi^2} \int_0^\infty k^3 \lambda_{\nu_\ell} \nabla f_{\nu_\ell}^{eq} dk. \tag{11}$$

To complete the calculation of $\mathbf{F}_{\nu_\ell}$ we need the mean free paths of the neutrinos or equivalently, the opacities κ_{ν_ℓ}.

The neutral current reactions equally affect all flavors. Since the energies available in the system are much smaller than the mass of the τ lepton, the ν_τ does not participate in the charge-current interactions. For the same reason, and for simplicity, we also ignore the presence of muons. Accordingly, the main sources of neutrino opacities are

$$\nu_\ell + n \rightleftarrows \nu_\ell + n \,, \tag{12}$$

$$\nu_\ell + p \rightleftarrows \nu_\ell + p \,, \tag{13}$$

$$\nu_e + n \rightleftarrows e^- + p \,. \tag{14}$$

In terms of the cross sections for these reactions the $\lambda_{\nu_\ell}^{-1}$ are given by

$$\lambda_{\nu_e}^{-1} = \lambda_{\nu_{\mu,\tau}}^{-1} + N_n \sigma_{abs}, \tag{15}$$

$$\lambda_{\nu_{\mu,\tau}}^{-1} = N_n \sigma_n + N_p \sigma_p \,, \tag{16}$$

where $N_{n,p}$ are the neutron and proton number densities and

$$\sigma_n = \frac{G_F^2}{4\pi} \left(1 + 3g_A^2\right) k^2 \,, \tag{17}$$

$$\sigma_p = \sigma_n \left[1 - \frac{8\sin^2\theta_W}{1 + 3g_A^2} \left(1 - 2\sin^2\theta_W\right)\right] \,, \tag{18}$$

$$\sigma_{abs} = 4\,\sigma_n \,. \tag{19}$$

Here, $\sin^2\theta_W \simeq 0.23$ and $g_A \simeq 1.26$ is the renormalization for the axial-vector current of the nucleons. In Eq. (19) we neglected contributions of order $(m_n - m_p)/k$, where m_n and m_p are the neutron and proton mass, respectively.

From the previous formulae we immediately see that

$$\lambda_{\nu_\ell}^{-1} = \kappa_{\nu_\ell}\rho = \varkappa_{\nu_\ell} k^2 \rho \,, \tag{20}$$

where the coefficients $\varkappa_{\nu_\ell}$ are constants. By making $m_n \cong m_p$ and taking the typical values $Y_n \approx 0.9$ and $Y_p \approx 0.1$ for the nucleon fractions $Y_{n,p} = N_{n,p}/(N_n + N_p)$, we get

$$\varkappa_{\nu_e} = 3 \times 10^{-25}\,\mathrm{MeV}^{-5}, \tag{21}$$

$$\varkappa_{\nu_{\mu,\tau}} = 0.7 \times 10^{-25}\,\mathrm{MeV}^{-5}. \tag{22}$$

When Eq. (20) is substituted in the integrand of Eq. (11), the momentum flux transported by neutrinos diffusing from the core to the exterior of the protostar becomes

$$\mathbf{F}_{\nu_\ell} = -\frac{1}{6\pi^2}\frac{1}{\varkappa_{\nu_\ell}\rho}\nabla\left(F_1[\eta_{\nu_\ell}]\,T^2\right) \,. \tag{23}$$

In what follows we assume that the neutrino chemical potentials are negligible, and using $F_1[0] = \pi^2/12$ and $F_3[0] = 7\pi^4/120$, from Eqs. (8) and (23), we obtain

$$U_{\nu_\ell} = \frac{7\pi^2}{240}\,T^4 \,, \tag{24}$$

$$\mathbf{F}_{\nu_\ell} = -\frac{1}{72\varkappa_{\nu_\ell}\rho}\,\nabla T^2 \,. \tag{25}$$

The last results will be used in Section 6, when we calculate the asymmetric neutrino emission in the context of a specific model for the protoneutron star.

4. Neutrino Oscillations in a Magnetized Medium

A notable consequence of neutrino mixing is the phenomena of neutrino oscillations established five decades ago by Pontecorvo[37, 38]. It corresponds to the periodic variation in the flavor content of a neutrino beam as it evolves from the production point. In the last years compelling evidence of neutrino oscillations has been obtained in experiments with atmospheric, solar, accelerator, and reactor neutrinos[39, 40]. The most plausible interpretation for these new phenomena is that neutrinos are massive particles and the weak-interaction states $|\nu_\ell\rangle$ ($\ell = e, \mu, \tau$) are linear combinations of the mass eigenstates $|\nu_j\rangle$ ($j = 1, 2, 3$), with masses m_j:

$$|\nu_\ell\rangle = U_{\ell j} \, |\nu_j\rangle, \tag{26}$$

where $U_{\ell j}$ are the elements of a unitary mixing matrix, the analogue of the CKM matrix in the quark sector. Consider a beam of ν_α created at a certain point with a 3-momentum $\mathbf{k}$. When the state given in Eq. (26) evolves, each of the mass eigenstates acquires a different phase $e^{-iE_j t}$, with $E_j = \sqrt{m_j^2 + k^2}$ ($k = |\mathbf{k}|$). As a consequence, at a certain distance $r \cong t$ from the source the neutrino beam will not correspond to pure ν_ℓ but turns partially into other flavors.

The pattern of neutrino oscillations is modified when neutrinos propagate through a material medium. The basic reason is that the energy-momentum relation of an active neutrino is affected by its coherent interactions with the particles that constitute the medium. These modifications are described in terms of a potential energy V_ℓ, which is related to the real part of the refraction index n_ℓ according to $\mathrm{Re}\, n_\ell = 1 - V_\ell/k$ and can be calculated from the background contributions to the neutrino self-energy[41]. If, like in normal matter, electrons but not heavier leptons are present in the background, then only ν_e scatter through charge current interactions, while all the flavors interact via the neutral current. Therefore, the refraction index differs for neutrinos of different flavors and they will acquire distinct phases when propagating with the same momentum through the medium. This fact can have impressive consequences in the case of mixed neutrinos and, under favorable conditions, flavor transformations are significantly enhanced in a medium with a varying density even when mixing in vacuum is small. This is the essence of the MSW mechanism[42, 38] whose large mixing realization provides the solution to the solar neutrino problem[40, 43].

Besides the modifications of the dispersion relation, neutrinos acquire in matter an effective coupling to the electromagnetic field by means of their weak interaction with the background particles[15]. As a result, in the presence of an external magnetic field there are additional contributions to the neutrino refraction index and flavor transformations are affected, but in a way that preserves quirality[44], contrary to what happens for neutrino oscillations driven by transition magnetic moments[45]. In what follows, we will discuss neutrino oscillations in a magnetized medium for the simplest situation of mixing between two neutrino species, that is, ν_ℓ and another active neutrino $\nu_{\ell'}$ or a hypothetical sterile neutrino ν_s. In any case we can write

$$\mathbb{U} = \begin{pmatrix} \cos\theta & \sin\theta \\ -\sin\theta & \cos\theta \end{pmatrix}, \tag{27}$$

where θ is the vacuum mixing angle, a parameter to be fixed from the experimental results.

The Dirac wave function for relativistic (mixed) neutrinos with momentum **k** propagating in matter in the presence of a static uniform magnetic field is well approximated as follows[44]:

$$\psi \cong e^{i\mathbf{k \cdot x}} \begin{pmatrix} 0 \\ \phi_- \end{pmatrix} \chi(t) , \tag{28}$$

where we use the Weyl representation, ϕ_- is the Pauli spinor of negative helicity ($\vec{\sigma} . \hat{\varkappa} \, \phi_- = -\phi_-$) and $\chi(t)$ is a flavor-space spinor that satisfies

$$i \frac{d\chi}{dt} = \mathbb{H}_B \chi . \tag{29}$$

The Hamiltonian $\mathbb{H}_B$ is a 2×2 matrix that, in the basis of the flavor states, takes the form

$$\mathbb{H}_B = k + \frac{\mathbb{M}^2}{2k} + \mathbb{V}, \tag{30}$$

with

$$\mathbb{M} = \mathbb{U} \begin{pmatrix} m_1 & 0 \\ 0 & m_2 \end{pmatrix} \mathbb{U}^\dagger \tag{31}$$

and $\mathbb{V} = \text{diag}(V_\ell, V_{\ell',s})$.

The effective potential for active neutrinos can be written as[15, 47]

$$V_\ell = b_\ell - c_\ell \, e \, \hat{\mathbf{k}} \cdot \mathbf{B} , \tag{32}$$

where $e > 0$ is the proton charge. The coefficients b_ℓ and c_ℓ depend on the properties of the thermal background. To order G_F the expression for b_ℓ is independent of B, but this is not true in general for c_ℓ. In our analysis we will work within the weak-field limit, i.e., $B \ll \mu_e^2/2e$[46], and ignore any possible dependence of this coefficient on the magnetic field. The effect of strong magnetic fields on the neutrino propagation in a medium have been considered by some authors[47]. For (active) antineutrinos the potentials change in sign, i.e., $\bar{V}_\ell = -V_\ell$, while in the case of sterile neutrinos, since they do not interact with the background, we have

$$V_s = \bar{V}_s = 0 . \tag{33}$$

Eq. (29) is the extension of the MSW equation to the situation we are considering, and was first derived in Ref. [44]. The components $\chi_e(t)$ and $\chi_{\mu,s}(t)$ of $\chi(t)$ give the amplitude to find the neutrino in the corresponding flavor state. For a uniform background $\chi(t) = \sum_{i=1,2} (\varsigma_i^\dagger \chi(0)) \varsigma_i e^{i\omega_i t}$, where $\omega_i(p)$ are the energies of the neutrino modes in the medium and the vectors $\varsigma_{1,2}$ are the solutions of the eigenvalue condition $\mathbb{H}_B \varsigma_i = \omega_i \varsigma_i$. For oscillations the relevant quantity is

$$\omega_2 - \omega_1 = \frac{1}{2k} \sqrt{\left(\Delta m^2 \cos 2\theta - 2k\mathcal{V} \right)^2 + \left(\Delta m^2 \sin 2\theta \right)^2} , \tag{34}$$

where $\mathcal{V}(r) = V_\ell(r) - V_{\ell',s}(r)$ and $\Delta m^2 \equiv m_2^2 - m_1^2$. When the medium is inhomogenous, along the neutrino path ($r \cong t$) the elements of $\mathbb{V}$ are functions of t, and $\chi(t)$ has to be determined by solving Eq. (29) with $\mathbb{H}_B$ a time dependent matrix. At each point r, $\mathbb{H}_B$ can be diagonalized by the unitary transformation

$$\mathbb{U}_m(r) = \begin{pmatrix} \cos\theta_m & \sin\theta_m \\ -\sin\theta_m & \cos\theta_m \end{pmatrix}, \tag{35}$$

with the mixing angle in matter $\theta_m(r)$ given by

$$\tan 2\theta_m = \frac{\Delta m^2 \sin 2\theta}{\Delta m^2 \cos 2\theta - 2k\mathcal{V}}. \tag{36}$$

A similar expression is valid in the case of antineutrinos with $\Delta m^2 \cos 2\theta + 2k\mathcal{V}$ in the denominator. In general $\mathcal{V}$ and consequently the mixing angle in matter depends on the magnetic field. As a function of $\mathcal{V}$, $\sin^2 2\theta_m$ exhibits the characteristic form of a Breit-Wigner resonance. For neutrinos (antineutrinos) with momentum $\mathbf{k}$ the resonance condition $(\sin^2 2\theta_m = 1)$ is

$$\mathcal{V}(r_R) = \pm \frac{\Delta m^2}{2k} \cos 2\theta, \tag{37}$$

where $r_R = r_R(\mathbf{k})$ and the sign $+ (-)$ corresponds to neutrinos (antineutrinos). Notice that for active-active oscillations the neutral current contributions to $\mathcal{V} = V_\ell - V_{\ell'}$ cancel since they are the same for both flavors. This is not true in the case of active-sterile oscillations where $\mathcal{V} = V_\ell$, and such contributions play a relevant role in the resonant transformations for $\nu_\ell \longleftrightarrow \nu_s$ (or $\bar{\nu}_\ell \longleftrightarrow \bar{\nu}_s$). The efficiency of the flavor transformation depends on the adiabaticity of the process. The conversion will be adiabatic when the oscillation length at the resonance is smaller than the width of the enhancement region. This imposes the condition

$$\left| \frac{d\mathcal{V}}{dr} \right|_{r_R} < \left(\frac{\Delta m^2}{2k} \sin 2\theta \right)^2. \tag{38}$$

The dispersion relation of neutrinos in a magnetized plasma made of electrons, protons, and nucleons have been calculated in Ref. [16], including the contributions due to the anomalous magnetic moment coupling of the nucleons to the photon. From the results given there we have

$$b_e = \sqrt{2}G_F \left[N_e + \left(\frac{1}{2} - 2\sin^2\theta_W \right)(N_p - N_e) - \frac{1}{2}N_n \right] + \tilde{b}_e, \tag{39}$$

$$b_{\mu,\tau} = \sqrt{2}G_F \left[\left(\frac{1}{2} - 2\sin^2\theta_W \right)(N_p - N_e) - \frac{1}{2}N_n \right] + \tilde{b}_{\mu,\tau}, \tag{40}$$

$$c_e = 2\sqrt{2}G_F \left[g_A \left(1 + 2m_p \frac{\kappa_p}{e} \right) C_p - g_A 2m_n \frac{\kappa_n}{e} C_n - C_e \right], \tag{41}$$

$$c_{\mu,\tau} = 2\sqrt{2}G_F \left[g_A \left(1 + 2m_p \frac{\kappa_p}{e} \right) C_p - g_A 2m_n \frac{\kappa_n}{e} C_n + C_e \right], \tag{42}$$

where $\kappa_{n,p}$ are the anomalous part of the nucleon magnetic moments given by

$$\kappa_n = -1.91 \frac{e}{2m_n}, \qquad \kappa_p = 1.79 \frac{e}{2m_p}, \tag{43}$$

and

$$\tilde{b}_\ell = \sqrt{2} G_F \left[N_{\nu_\ell} - N_{\bar{\nu}_\ell} + \sum_{\ell'} \left(N_{\nu_{\ell'}} - N_{\bar{\nu}_{\ell'}} \right) \right] , \tag{44}$$

$$C_\flat = \frac{1}{2} \int \frac{d^3 p}{(2\pi)^3} \frac{1}{2E} \frac{d}{dE} f_\flat \quad (\flat = e, n, p) . \tag{45}$$

In these equations N_e and $N_{\nu_\ell} (N_{\bar{\nu}_\ell})$ are the number densities of electrons and neutrinos (antineutrinos) of flavor ℓ, respectively and $f_\flat$ are the distribution functions of the electrons, protons, and neutrons.

We model the atmosphere of a protoneutron star as a neutrino gas diffusing through a plasma made of relativistic degenerate electrons and classical nonrelativistic nucleons. In such conditions $C_e = -\mu_e/8\pi^2$ and $C_{n,p} = -N_{n,p}/8Tm_{n,p}$[16, 46], where T is the background temperature and $\mu_e = \left(3\pi^2 N_e \right)^{1/3}$ is the electron chemical potential. The neutrino fractions are much smaller than the electron fraction $Y_e = N_e/(N_n + N_p) \sim 0.1$ and, in the context of our analysis, the contributions to b_{ν_ℓ} coming from the neutrino-neutrino interactions (denoted by $\tilde{b}_\ell$) can be neglected. With this approximation in mind, putting $m_p \simeq m_n$, and taking into account that by electrical neutrality $N_e = N_p$, the coefficients in the effective potentials become:

$$b_e \simeq -\frac{G_F}{\sqrt{2}} (1 - 3Y_e) \frac{\rho}{m_n} , \tag{46}$$

$$b_{\mu,\tau} \simeq -\frac{G_F}{\sqrt{2}} (1 - Y_e) \frac{\rho}{m_n} , \tag{47}$$

$$c_e \simeq -\sqrt{2} G_F \left(\frac{3 + Y_e}{5Tm_n^2} \rho - \frac{\mu_e}{4\pi^2} \right) , \tag{48}$$

$$c_{\mu,\tau} \simeq -\sqrt{2} G_F \left(\frac{3 + Y_e}{5Tm_n^2} \rho + \frac{\mu_e}{4\pi^2} \right) . \tag{49}$$

It should be noticed that in Ref. [20] c_e vanishes because all the components of the stellar plasma are assumed to be degenerated, and the contributions due to the anomalous magnetic moment of the nucleons are not included.

We now use the effective potentials V_ℓ with b_ℓ and c_ℓ given by Eqs. (46)-(49) to examine the magnetic field effect on the resonant neutrino oscillations within a protoneutron star. In a linear approximation, we can write

$$r_R(\mathbf{k}) = r_o(k) + \delta(k) \cos \alpha , \tag{50}$$

where r_o and δ are ordinary functions of the magnitude of the neutrino momentum and α is the angle between $\mathbf{k}$ and $\mathbf{B}$. For a certain value of k, Eq. (50) determines a spherical shell limited by the spheres of radii $r_o(k) \pm \delta(k)$, where the resonance condition is verified for neutrinos (or antineutrinos) moving in different directions with respect to the magnetic field. The quantity r_o corresponds to the radius of the resonance sphere when $B = 0$ and in the case of oscillations between ν_e and ν_μ (or ν_τ) is given by

$$G_F \sqrt{2} Y_e \frac{\rho}{m_n} \bigg|_{r_o} = \pm \frac{\Delta m^2}{2k} \cos 2\theta . \tag{51}$$

The left-hand side corresponds to the difference $b_e - b_{\mu,\tau}$ evaluated at $r = r_o$ and is positive definite. Therefore, assuming $\Delta m^2 \cos 2\theta > 0$, the above condition can be satisfied for $\nu_e \longleftrightarrow \nu_{\mu,\tau}$ but not for $\bar{\nu}_e \longleftrightarrow \bar{\nu}_{\mu,\tau}$. To be effective as a kick mechanism the resonance transformations of active neutrinos have to take place in the region between the electron and muon (tau) neutrinospheres. In this case, the ν_e are trapped by the medium, but the $\nu_{\mu,\tau}$ produced this way are above their neutrinosphere and, when moving outside, can freely escape from the star. In the presence of a magnetic field, the emission points for $\nu_{\mu,\tau}$ having the same k but different directions are not at the same radius, which originates an asymmetry in the momentum they carried away.

Replacing k in Eq. (51) by the thermal average $\langle k \rangle \simeq 3.15\, T$ and taking into account that $\rho \sim 10^{11}\,\mathrm{g\,cm^3}$ and $T \sim 4\,\mathrm{MeV}$ at R_{ν_e}, the requirement of having the resonance within the electron neutrinosphere implies $\Delta m^2 \cos 2\theta \gtrsim 2 \times 10^4\,\mathrm{eV^2}$, which is excluded by the experimental results on neutrino oscillations and the cosmological limits. For this reason, in what follows we concentrate only on the active-sterile oscillations. In this case, the resonance condition when $B = 0$ takes the form

$$b_\ell(r_{o\ell}) = \mp \frac{\Delta m^2}{2k} \cos 2\theta \,, \tag{52}$$

for $\nu_\ell \longleftrightarrow \nu_s$ ($\bar{\nu}_\ell \longleftrightarrow \bar{\nu}_s$). In the static model we use for the protoneutron star, a good approximation is to take $Y_e < 1/3$. Therefore, if $\Delta m^2 \cos 2\theta > 0$, the resonant condition is verified only by antineutrinos and from now on we restrict ourselves to this situation.

To determine the quantity δ in Eq. (50) we substitute V_ℓ as given by Eq. (32) in Eq. (37) and expand b_ℓ up to first order in δ_ℓ. Proceeding in this way we get

$$\delta_\ell(k) = \mathcal{D}_\ell(k)\, eB \,, \tag{53}$$

with

$$\mathcal{D}_\ell(k) = \frac{1}{h_{b_\ell}^{-1}} \left. \frac{c_\ell}{b_\ell} \right|_{r_{o\ell}} , \tag{54}$$

where we have defined $h_g^{-1} \equiv \frac{d}{dr} \ln g$ for any function $g(r)$. Using Eqs. (46)-(49) we find explicitly

$$\mathcal{D}_e = \frac{2}{(1 - 3Y_e)\, h_\rho^{-1} - 3Y_e h_{Y_e}^{-1}} \left. \left(\frac{3 + Y_e}{5Tm_n} - \frac{m_n \mu_e}{4\pi^2 \rho} \right) \right|_{r_{oe}} , \tag{55}$$

$$\mathcal{D}_{\mu,\tau} = \frac{2}{(1 - Y_e)\, h_\rho^{-1} - Y_e h_{Y_e}^{-1}} \left. \left(\frac{3 + Y_e}{5Tm_n} + \frac{m_n \mu_e}{4\pi^2 \rho} \right) \right|_{r_{o\mu,\tau}} . \tag{56}$$

Finally from Eq. (38), discarding terms proportional to the magnetic field, the following restriction to the vacuum mixing angle results

$$\tan^2 2\theta > \left. \left| \frac{h_{b_\ell}^{-1}}{b_\ell} \right| \right|_{r_{o\ell}} , \tag{57}$$

which guarantees that the evolution will be adiabatic and an almost complete flavor transformation for small values of θ.

5.   Neutrino Momentum Asymmetry

If the temperature and density profiles are isotropic, then the momentum flux in the protoneutron star atmosphere is radial (see Eq. (25)). As a consequence, in most of the works on the kick mechanism driven by matter neutrino oscillations the resonance condition was evaluated assuming that neutrinos move outside with a momentum $\langle k \rangle = (7\pi^4/180\,\zeta(3))\,T \simeq 3.15\,T$ pointing in the radial direction. In presence of a magnetic field this approach leads to the concept of a deformed resonance surface, which acts as an effective neutrinosphere. However, at each point within the neutrinosphere there are neutrinos with momentum **k** pointing in every directions and the resonance condition as given by Eq. (37) actually defines a spherical surface for each value of $\cos\alpha$. This surface acts as a source of sterile neutrinos produced through the resonant $\bar{\nu}_\ell$ transformations, which move in the $\hat{\mathbf{k}}$ direction. Therefore, as was mentioned above, for a given k the resonant transformations take place within a spherical shell, and not on a deformed spherical surface. A more careful calculation of the pulsar kick along this line has been carried out in Ref. [22] and here we follow the same approach.

Let us consider the situation where the resonance shell is totally contained within the $\bar{\nu}_\ell$ neutrinosphere. The sterile neutrinos produced there that move out freely escape from the protostar, but those directed toward the inside cross the resonance surface again and reconvert into $\bar{\nu}_\ell$ which, being within their own neutrinosphere, are thermalized. Consequently, only those $\bar{\nu}_s$ going outward can leave the star and the resonance surfaces for different momentum directions behave as effective emission semispheres of sterile antineutrinos. For **k** pointing in opposite directions the radii of the respective semispheres are $r_R^\pm = r_o(k) \pm \delta(k)\,|\cos\alpha|$, and therefore they have different areas that generate a difference in the momentum carried away by the sterile neutrinos leaving the star in opposite directions. There also exists a temperature variation within the resonance shell such that the temperatures of the emission semispheres for $\bar{\nu}_s$ going outside in opposite directions are different $(T(r_R^-) > T(r_R^+))$. The variations in the area and the temperature tend to compensate each other when the contributions of $\bar{\nu}_s$ emitted in every direction are added up. In the calculation below we take both semispheres into account when explicitly computing the asymmetry in the total momentum $\mathcal{K}$ emitted by the cooling protoneutron star. In this way, we improve the calculation of Ref. [22] where the temperature within the resonance shell was assumed to be uniform.

The active neutrinos and antineutrinos are emitted isotropically from their respective neutrinospheres and do not contribute to the momentum asymmetry $\Delta\mathcal{K}$. Therefore, the only nonvanishing contribution can come from the $\bar{\nu}_s$. If their emission lasts for an interval Δt of the order of a few seconds, then in the static model we are using for the protostar $\Delta\mathcal{K} = |K_B|\,\Delta t$, where K_B is the component along the magnetic field direction of the neutrino momentum emitted per time unit. According to Eq. (11) we can write

$$K_B = \frac{1}{(2\pi)^2}\int_0^\infty k^3 dk \int_0^\pi \sin\theta d\theta \int_0^{\frac{\pi}{2}} \sin\theta' d\theta' \int_0^{2\pi} d\phi' \, r_{R_\ell}^2(\mathbf{k}) f_{\bar{\nu}_s}\,\hat{\mathbf{k}}\cdot\hat{\mathbf{B}}\,, \qquad (58)$$

where $\hat{\mathbf{k}}\cdot\hat{\mathbf{B}} = \cos\theta\cos\theta' - \sin\theta\sin\theta'\sin\phi'$ and the distribution function for sterile neutrinos $f_{\bar{\nu}_s}$ is evaluated at r_{R_ℓ}. We have chosen a reference frame $(x,\,y,\,z)$ fixed to the protostar where the magnetic field coincides with the z-axis, $\mathbf{B} = B\,\hat{\mathbf{z}}$. In addition, at each point within the resonance shell we use the local frame introduced in Section 2 to evaluate the components of the energy-momentum tensor. Here, the upper limit for the integration on θ' is $\pi/2$ instead of π, because we have to include only the contributions from the sterile neutrinos going outside.

To proceed further with the calculation of K_B we need to know $f_{\bar{\nu}_s}$. The proper treatment of the interplay of collisions and oscillations for trapped neutrinos requires the use of the density matrix. For our purpose a less elaborate description will suffice and we put $f_{\bar{\nu}_s} \cong \mathcal{P}(\bar{\nu}_\ell \to \bar{\nu}_s) f_{\bar{\nu}_\ell}$, where $\mathcal{P}(\bar{\nu}_\ell \to \bar{\nu}_s)$ is the probability for $\bar{\nu}_\ell$ to convert into $\bar{\nu}_s$. Let us assume that the adiabaticity condition given by Eq. (57) is verified in the interior of the $\bar{\nu}_\ell$ neutrinosphere, and hence for small mixing $\mathcal{P}(\bar{\nu}_\ell \to \bar{\nu}_s) \cong 1$. Accordingly in Eq. (58) we take $f_{\bar{\nu}_s}(r_R) \simeq f_{\bar{\nu}_\ell}(r_R)$, with

$$
f_{\bar{\nu}_\ell}(r_R) \cong f_{\bar{\nu}_\ell}^{eq}(r_{o\ell}) - \left(\frac{\cos\theta'}{\varkappa_{\bar{\nu}_\ell} k^2 \rho} \frac{df_{\bar{\nu}_\ell}^{eq}}{dr} - \hat{\mathbf{k}} \cdot \hat{\mathbf{B}} \frac{df_{\bar{\nu}_\ell}^{eq}}{dr} \delta_\ell(k) \right)\Bigg|_{r_{o\ell}}, \tag{59}
$$

where a term proportional to the derivative of $\rho^{-1} df_{\bar{\nu}_\ell}^{eq}/dr$ has been neglected, in agreement with the diffusive approximation we are employing. Substituting the expansion (59) into Eq. (58) and keeping terms that are at most linear in δ_ℓ we find

$$
\begin{aligned}
K_B \cong \frac{2}{3\pi} \int_0^\infty & dk \, k^3 r_{o\ell}(k) \, \delta_\ell(k) \\
& \left(f_{\bar{\nu}_\ell}^{eq} - \frac{1}{2\varkappa_{\bar{\nu}_\ell} k^2 \rho} \frac{df_{\bar{\nu}_\ell}^{eq}}{dr} + \frac{1}{2} r_{o\ell}(k) \frac{df_{\bar{\nu}_\ell}^{eq}}{dr} \delta_\ell(k) \right)\Bigg|_{r_{o\ell}}.
\end{aligned} \tag{60}
$$

The energy released during the core collapse of the progenitor star is approximately equipartitioned among all the neutrino and antineutrino flavors. In the context of our analysis this means that $K_r \Delta t \simeq \mathcal{K}/6$, where K_r represents the momentum rate of the sterile neutrinos in the radial direction and is given by an expression similar to Eq. (58) with $\hat{\mathbf{k}} \cdot \hat{\mathbf{B}}$ replaced by $\hat{\mathbf{k}} \cdot \hat{\mathbf{r}}$. Following the same procedure outlined above we find

$$
K_r \cong \frac{1}{2\pi} \int_0^\infty dk \, k^3 \, r_{o\ell}^2(k) \left(f_{\bar{\nu}_\ell}^{eq} - \frac{2}{3\varkappa_{\bar{\nu}_\ell} \rho k^2} \frac{df_{\bar{\nu}_\ell}^{eq}}{dr} \right)\Bigg|_{r_{o\ell}}. \tag{61}
$$

In terms of K_B and K_r the fractional asymmetry in the total momentum neutrinos carry away becomes

$$
\frac{\Delta \mathcal{K}}{\mathcal{K}} = \frac{|K_B|}{6 K_r}. \tag{62}
$$

To evaluate the remaining integrals in Eqs. (61) and (60) the explicit dependence on k of the functions $r_{o\ell}$ and $\mathcal{D}_\ell$ have to be known. Simple analytical results can be derived by replacing these functions by the constant quantity $\bar{r}_{o\ell}$ and $\overline{\mathcal{D}}_\ell = \mathcal{D}_\ell(\bar{r}_{o\ell})$, where $\bar{r}_{o\ell}$ is determined from the resonance condition when k is replaced by its thermal average $3.15\,T(r)$:

$$
6.3\,T(\bar{r}_{o\ell})\,b_\ell(\bar{r}_{o\ell}) = \Delta m^2 \cos 2\theta. \tag{63}
$$

Proceeding in this manner we get

$$K_r \cong \frac{\pi \bar{r}_{o\ell}^2}{48} \left(\frac{7\pi^2 T^4}{5} - \frac{4}{3\kappa_{\bar{\nu}_\ell} \rho} \frac{dT^2}{dr} \right) \Bigg|_{\bar{r}_{o\ell}} , \tag{64}$$

$$K_B \cong \frac{\pi \bar{r}_{o\ell} \bar{\delta}_\ell}{36} \left(\frac{7\pi^2 T^4}{5} + \frac{7\pi^2 \bar{r}_{o\ell}}{10} \frac{dT^4}{dr} - \frac{1}{\varkappa_{\bar{\nu}_\ell} \rho} \frac{dT^2}{dr} \right) \Bigg|_{\bar{r}_{o\ell}} . \tag{65}$$

In these expressions the terms that depend on dT^2/dr come from the diffusive part of the distribution function. In the regime considered here they are smaller than the other contributions and in what follows we will neglect them.

Suppose that the pulsar momentum K_{pul} is entirely due to the active-sterile oscillation driven kick, then $\Delta K = K_{pul}$. For a pulsar with a mass $M \approx M_\odot$ and a translational velocity $v \approx 500 \, \mathrm{km \, s}^{-1}$ we have $K_{pul} \approx 10^{41} \, \mathrm{g \, cm \, s}^{-1}$. Since neutrinos are relativistic, the total amount of momentum they carry is equal to the gravitational energy liberated by means of them, that is $K \approx 10^{43} \, \mathrm{g \, cm \, s}^{-1}$. Therefore, as mentioned in the Introduction, $\Delta K / K$ must be of the order of 0.01 to obtain the required kick. In this case Eq. (62) yields

$$eB \cong \frac{0.045 \, \bar{r}_{o\ell}}{\left| (1 + 2\bar{r}_{o\ell} \bar{h}_T^{-1}) \, \overline{\mathcal{D}}_\ell \right|} , \tag{66}$$

with $\bar{h}_T^{-1}$ denoting the logarithmic derivative of $T(r)$ evaluated at $\bar{r}_{o\ell}$. This formula allows us to determine the magnitude of the magnetic field once the temperature, the baryon density, and the electron fraction profiles are known. Below we do it for the $\bar{\nu}_\mu - \bar{\nu}_s$ conversion under the assumption that the resonance layer is entirely within the $\bar{\nu}_\mu$-sphere. However, we will first indicate how $Y_e(r)$ is determined by means of $T(r)$ and $\rho(r)$ within our approximate description of a protoneutron star.

For $\mu_{\nu_e} \simeq 0$, the β equilibrium gives the relation $\mu_n - \mu_p \simeq \mu_e$ among the chemical potentials of the background fermions. Hence, for non-relativistic nucleons $N_n/N_p \simeq \exp[-(m_n - m_p - \mu_e)/T]$ and

$$Y_e \simeq \frac{1}{1 + e^{\mu_e/T}} , \tag{67}$$

where we used the fact that in the condition prevailing in a protoneutron star μ_e is much larger than the difference between the neutron and the proton masses. Since electrons are degenerated $\mu_e = \left(3\pi^2 \rho Y_e / m_n \right)^{1/3}$ and Eq. (67) is an implicit equation for Y_e to be solved numerically once ρ and T are known as a function of r. Taking the derivative of Eq. (67) with respect to r, $h_{Y_e}^{-1}$ can be expressed in terms of h_ρ^{-1} and h_T^{-1} as follows

$$h_{Y_e}^{-1} = - \left(h_\rho^{-1} - 3h_T^{-1} \right) \frac{1 - Y_e}{1 - Y_e + \frac{3T}{\mu_e}} . \tag{68}$$

To estimate B we now assume that the density and the temperature are related according to $\rho = \rho_c (T/T_c)^3$[33], where ρ_c and T_c are the respective values of these quantities at the core radius R_c. From this expression we see that $h_\rho^{-1} = 3 \, h_T^{-1}$ which implies that

$h_{Y_e}^{-1} = 0$ (see Eq. (68)). Therefore, the electron fraction Y_e and as a consequence the ratio μ_e / T are constants. As core parameters we take $R_c = 10\,\text{km}$, $\rho_c = 10^{14}\,\text{g cm}^{-3}$, and $T_c = 40\,\text{MeV}$, while for the density profile we adopt the potential law $\rho(r) = \rho_c (R_c/r)^4$ which yields $h_\rho^{-1} = -4/r$ and $T(r) = T_c (R_c/r)^{4/3}$. The $\bar{\nu}_\mu$-sphere is taken stationary at a density of about $10^{12}\,\text{g cm}^{-3}$ that corresponds to a radius $R_{\bar{\nu}_\mu} \simeq 3\,R_c$ and a temperature $T_{R_{\bar{\nu}_\mu}} \simeq 9\,\text{MeV}$. Using these results as well as the formula for $\mathcal{D}_\mu$ given in Eq. (56), from Eq. (66) we obtain the following simple result:

$$B = 5 \times 10^{17} \left(1 + 0.3 \frac{T_c}{\bar{T}} \right)^{-1} \frac{\bar{T}}{T_c}\,\text{G}, \qquad (69)$$

where $\bar{T} = T(\bar{r}_{o_\mu})$, with $R_c \leq \bar{r}_{o_\mu} \leq R_{\bar{\nu}_\mu}$. We see that the required intensity of the magnetic field decreases monotonically from $3.8 \times 10^{17}\,\text{G}$ at the core radius to $4.8 \times 10^{16}\,\text{G}$ at the $\bar{\nu}_\mu$-sphere. According to Eqs. (52) and (47), within the specified interval for $\bar{r}_{o_\mu}$ the range of the allowed values of the oscillation parameters are $7 \times 10^8\,\text{eV}^2 \gtrsim \Delta m^2 \cos 2\theta \gtrsim 10^5\,\text{eV}^2$, and for small mixing we have $25\,\text{KeV} \gtrsim m_s \gtrsim 0.3\,\text{KeV}$. The requirement that the resonant conversion be adiabatic imposes the condition $\tan^2 2\theta \gtrsim 3 \times 10^{-11} \left(\bar{r}_{o_\mu}/R_c \right)^3$ on the mixing angle. A last comment is in order, the ratio of the diffusive term to the isotropic one in Eqs. (64) and (65) can be expressed as $10^{-4} \left(\bar{r}_{o_\mu}/R_c \right)^{17/3}$, which illustrates the validity of the approximation done in writing Eq. (66). In the next section we repeat these calculations in the realm of the spherical Eddington model, which provides a simple but quite physically reasonable description of a static neutrino atmosphere.

6. Spherical Eddington Model

The actual configuration of the neutrinosphere and the resonance region depend on details of the protoneutron star, such as the density, temperature, pressure, and leptonic fraction profiles. The asymmetry in the momentum flux also depends on these features and on the configuration of the magnetic field. The present knowledge about protoneutron stars is not enough to single out a definite model for their structure, but there are some well established general characteristics that help towards proposing models that are both workable and plausible. An important feature is that the size and the shape of the different profiles do not suffer sudden changes during the time interval when the main neutrino emission takes place, i.e. $0.5\,s \lesssim t \lesssim 10\,s$. Thus, as a first approximation we can consider a stationary regime that greatly simplifies calculations. Another characteristic to be taken into account is that the bulk of neutrinos is produced in the inner core. Consequently, we assume that there is no neutrino production outside the core and hence the flux of diffusing neutrinos is conserved in this region. This flux is responsible for the transport of the energy liberated by the protostar.

As mentioned before, the system is supposed to be constituted by nucleons, with a density ρ, electrons, neutrinos and antineutrinos. Neutrinos are in thermal equilibrium with nucleons, satisfying a transport regime consistent with the diffusion approximation, and are assumed to have a vanishing chemical potential, $\mu_\nu = \mu_{\bar{\nu}} = 0$. The total energy density, the radial energy flux, and the pressure for neutrinos and antineutrinos in the case of spherical symmetry, can be obtained from Eqs. (24), (25) and (10)

$$U = \sum_\ell U_{\nu_\ell} = \frac{7}{40}\pi^2 T^4, \tag{70}$$

$$F = \sum_\ell F_{\nu_\ell} = -\frac{1}{36\bar{\varkappa}\rho}\frac{dT^2}{dr}, \tag{71}$$

$$P_\nu = \sum_\ell T_{\nu_\ell}^{ii} = \frac{U}{3}, \tag{72}$$

where $\bar{\varkappa}^{-1} = \varkappa_{\nu_e}^{-1} + 2\varkappa_{\nu_{\mu,\tau}}^{-1}$. The main contributions to the neutrino opacities were discussed in Section III. Using the results given there for $\varkappa_{\nu_e}$ and $\varkappa_{\nu_{\mu,\tau}}$ (Eqs. (21) and (22)) we obtain $\bar{\varkappa} = 3 \times 10^{-26}$ MeV^{-5}.

If in addition we assume that baryons constitute a nonrelativistic ideal gas, we have the Eddington model. It is simple and physically well justified, allowing a detailed discussion of the relevant characteristics of the protostar and the geometry of the resonance region. Originally proposed to describe a stellar photosphere, this model was adapted by Schinder and Shapiro[48] to a neutrino atmosphere with a plane geometry. The more realistic case of a spherical atmosphere was considered in Ref. [19] to analyze the geometrical effect on the asymmetric neutrino emission induced by resonant oscillations. Photons and electrons are of course present and, in fact, electrons make the leading contribution to the effective potential in the case of oscillations among active neutrinos. However, we can ignore both photons and electrons for the hydrodynamical description of the system.

The state equation that defines the system is given by the sum of the contributions from nucleons and neutrinos to the pressure gas, which is

$$P = \frac{\rho T}{m_n} + \frac{U}{3}. \tag{73}$$

In the Newtonian limit for the metric ($g_{oo} = -1 - 2\phi$, $g_{oj} = 0$, and $g_{ij} = \delta_{ij}$, with $\phi = -GM(r)/r$) the conservation of the total (neutrinos + matter) energy-momentum tensor yields the energy flux conservation

$$\frac{\partial(r^2 F)}{\partial r} = 0 \tag{74}$$

and the hydrostatic equilibrium equation

$$\frac{dP}{dr} = -(P + \rho + U)\frac{GM}{r^2}. \tag{75}$$

Here, $M(r) = 4\pi \int_0^r dr' r'^2 \rho(r')$ is the mass enclosed up to a distance r from the center.

Through the atmosphere the baryon density is $\rho \simeq (10^{11} - 10^{14})$ g cm^{-3} and $T \simeq (4 - 40)$ MeV, and thus we have $U \simeq (10^{-3} - 10^{-2})\rho$ and $P \simeq (4 \times 10^{-3} - 4 \times 10^{-2})\rho$. This means that on the right-hand side of Eqs. (73) and (75) we can ignore the contributions coming from the energy density of neutrinos and the baryonic density. In addition, Eq. (74) implies that $F = L_c/4\pi r^2$, where L_c is the luminosity of the protostar. Taking all the above into account we arrive at the following set of equations for an isotropic atmosphere of neutrinos in thermal equilibrium with an ideal gas of nucleons:

$$\frac{dT^2}{dr} = -\rho \frac{9\bar{\varkappa} L_c}{\pi r^2} , \tag{76}$$

$$P = \frac{\rho}{m_n} T, \tag{77}$$

$$\frac{dP}{dr} = -\rho \frac{GM}{r^2} , \tag{78}$$

$$\frac{dM}{dr} = 4\pi r^2 \rho . \tag{79}$$

Solving this system of equations yields the profiles of four functions throughout the atmosphere: pressure $P(r)$, temperature $T(r)$, baryonic density $\rho(r)$, and the enclosed mass $M(r)$. From them any other functions of interest can be calculated. To analyze the system it is convenient to introduce adimensional variables normalized to the corresponding values at the core: $x = r/R_c$, $m(x) = M/M_c$, $t(x) = T/T_c$, $\varrho(x) = \rho/\rho_c$, and $p(x) = P/P_c$. Proceeding in this way, Eqs. (76)-(79) can be rewritten

$$\frac{dt}{dx} = -b_c \frac{\varrho}{tx^2} , \tag{80}$$

$$p = t\varrho , \tag{81}$$

$$\frac{dp}{dx} = -c_c \frac{m\varrho}{x^2} , \tag{82}$$

$$\frac{dm}{dx} = d_c x^2 \varrho , \tag{83}$$

where

$$b_c = \frac{9\bar{\varkappa}\rho_c L_c}{2\pi R_c T_c^2} , \qquad c_c = \frac{Gm_n M_c}{R_c T_c} , \qquad d_c = \frac{4\pi\rho_c R_c^3}{M_c} . \tag{84}$$

are adimensional constants. Thus the set of functions that solve the previous system of equations depends only on these three combinations of the core parameters.

By eliminating ϱ in Eqs. (80) and (82) we arrive to

$$\frac{dp}{dx} = \frac{c_c}{b_c} mt \frac{dt}{dx} . \tag{85}$$

The last equation is easily integrated as follows:

$$p = \frac{t^2 - a}{1 - a} , \tag{86}$$

where

$$a(x) = 1 - \frac{2b_c}{c_c \bar{m}(x)} \tag{87}$$

and $\bar{m}(x)$ is an effective mass defined by

$$\int_1^x dx' \, m(x') \, t \frac{dt}{dx'} \equiv \bar{m}(x) \int_1^x dx' \, t \frac{dt}{dx'} . \tag{88}$$

From Eqs. (81) and (86) we can also express the density in terms of the temperature as

$$\varrho = \frac{t^2 - a}{t(1-a)} ,$$

(89)

which, together with Eq. (80), yields a first order differential equation for t

$$\frac{dt}{dx} + \frac{b_c}{t^2 x^2} \frac{t^2 - a}{1-a} = 0 .$$

(90)

At $x = 1$ ($R = R_c$) we have $dt/dx|_{x=1} = -b_c$, which is independent of $a(x)$. For an idealized atmosphere, where the model would apply to the whole space, the physical solutions would correspond to an infinite protostar where the temperature has an asymptotic behavior for $x \gg 1$, such that the adimensional temperature tends to $t_s \simeq \sqrt{a}$. Thus, the function $a(x)$ varies in the range $1 - 2b_c/c_c < a < t_s^2$ for $1 < x < \infty$.

Eq. (90) has no analytical solution when a is a function of x. An approximate solution can be found by the following procedure. Let us consider the differential equation (90) with a constant. Then, an analytical (implicit) solution is given by

$$t - 1 + \frac{\sqrt{a}}{2} \left[\ln\left(\frac{t - \sqrt{a}}{t + \sqrt{a}}\right) - \ln\left(\frac{1 - \sqrt{a}}{1 + \sqrt{a}}\right) \right] = \frac{b_c}{1-a} \left(\frac{1}{x} - 1\right) .$$

(91)

with $1 > t > \sqrt{a}$. If we replace the constant a by a well behaved function of x, the above expression still satisfies Eq. (90) at $x = 1$. For an infinite atmosphere, a good approximation to the exact solution is given by Eq. (91), with a now a function of $t(x)$:

$$a(t) = 1 - \frac{2b_c}{c_c} - A(1 - t) ,$$

(92)

where $A = \left(1 - \frac{2b_c}{c_c} - t_s^2\right) / (1 - t_s)$. From Eq. (92), we see that $a(t_s) = t_s^2$ and $a(1) = 1 - 2b_c/c_c$, which means that this ansatz fits the extreme values of $a(x)$.

Once ρ and T are known, we can find the electron fraction Y_e as a function of r by solving the implicit expression derived in Eq. (67). To examine the prediction of the model in connection with the kick mechanism, we consider a protoneutron star with reasonable values for its core parameters, the same as in Ref. [19]. Thus we take $M_c = M_\odot = 1.989 \times 10^{30}\,\text{kg}$, $R_c = 10\,\text{km}$, $L_c = 2 \times 10^{52}\,\text{erg s}^{-1}$, $\rho_c = 10^{14}\,\text{g cm}^{-3}$, and $T_c = 40\,\text{MeV}$. In this case,

$$b_c = 1.96 , \qquad c_c = 3.52 , \qquad d_c = 0.625 .$$

(93)

A numerical estimation for the asymptotic value of the solution of Eqs. (80)-(83) gives $t_s = 0.12$. The electron fraction takes the value $Y_e = 0.08$ at the core.

According to Eq. (20) in terms of the adimensional functions the mean free path for the electron neutrino becomes

$$\lambda_{\nu_e} \simeq \frac{8}{\varrho \, t^2}\,\text{cm} .$$

(94)

Inserting the above expression in the integrand of Eq. (1) and evaluating the resulting integral numerically we obtain $x_{\nu_e} \simeq 2.6$ for the radius of the ν_e-sphere. This yields $\rho(R_{\nu_e}) \simeq 1.3 \times 10^{11}\,\text{g cm}^{-3}$, $T(R_{\nu_e}) \simeq T_s = 4.8\,\text{MeV}$, and $Y_e(R_{\nu_e}) \simeq 0.1$, which are in

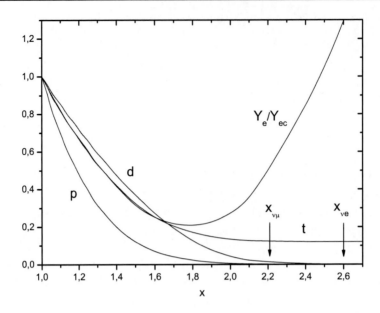

Figure 1. The different profiles normalized to the corresponding value at the core for the Eddington model discussed in Section VI.

reasonable agreement with typical values from numerical simulations[32, 4]. In the same way, for the muon (tau) neutrino we get $x_{\nu_{\mu,\tau}} \simeq 2.2$. The profiles that correspond to the solution of this model are given in Fig. 1.

From Eq. (66) we can estimate again the intensity of the magnetic field required for the kick in the case of $\bar{\nu}_\mu \longleftrightarrow \bar{\nu}_s$. Besides the temperature, density, and electron fraction profiles we also need their logarithmic derivatives. For T and ρ they are immediately calculated from Eqs. (80)-(82), and the corresponding results are

$$h_T^{-1} = -b_c \frac{\varrho}{tx} \frac{1}{r}, \tag{95}$$

$$h_\rho^{-1} + h_T^{-1} = -c_c \frac{m}{tx} \frac{1}{r}. \tag{96}$$

The formula for $h_{Y_e}^{-1}$ is obtained substituting the above expressions into Eq. (68). Now we have all the quantities needed to determine the magnetic field from Eq. (66) in the spherical Eddington model. The result is plotted in Fig. 2. From the curve presented in this figure we see that B decreases from a value of $8 \times 10^{16}\, G$ in the core up to a minimum of $3.8 \times 10^{16}\, G$ at $r_{o_\mu} \simeq 16\,\text{km}$. From this point the magnetic field increases steadily as we approach the surface of the neutrinosphere. In contrast with the model used at the end of Section 5, here at $r \simeq 21\, km$ the contributions from the geometrical and the temperature variations compensate each other. At this radius the factor $(1 + 2\bar{r}_{o_\ell})$ in the denominator of Eq. (66) vanishes and the magnetic field takes arbitrarily large values. In other words, no kick can be generated by this mechanism in the regions near the surface of the ν_μ-sphere.

In the region between the core and the surface of the neutrinosphere $R_c \leq r_{o_\mu} \leq R_{\nu_\mu}$,

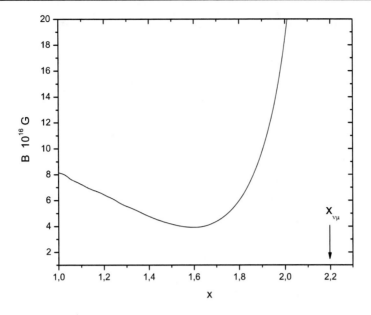

Figure 2. The magnetic field B required to produce the kick by the $\bar{\nu}_\mu \rightarrow \bar{\nu}_s$ resonant conversion for the Eddington model discussed in Section VI, normalized to the field required at the core, B_c.

the range of values allowed for the oscillation parameters is

$$10^9 \text{eV}^2 \gtrsim \Delta m^2 \cos 2\theta \gtrsim 10^6 \text{eV}^2, \tag{97}$$

For a small mixing angle this inequality translates into the condition $30\,\text{KeV} \gtrsim m_s \gtrsim 1\,\text{KeV}$ for the sterile neutrino mass. Small mixing is allowed by the adiabaticity condition, which warrants an efficient flavor neutrino conversion for $\tan^2 2\theta > 7 \times 10^{-12}$ at $r_{o_\mu} = R_c$ and $\tan^2 2\theta > 3 \times 10^{-9}$ at $r_{o_\mu} = R_{\nu_\mu}$. The results for m_s and the mixing angle agree with those derived in the previous section and are compatible with a sterile neutrino that is a viable dark matter candidate[49].

7.   Conclusion

In this work we have examined in detail a possible explanation of the large drift velocities observed in pulsars in terms of an asymmetric neutrino emission. The asymmetry is a consequence of the resonant neutrino conversions affected by the strong magnetic field characteristic of protoneutron stars. We have shown that for active-sterile neutrino oscillations this is a feasible kick mechanism. Two conditions must be simultaneously satisfied to generate a natal kick by this mechanism. First, the conversion has to take place at the interior of the neutrinosphere of the active neutrino, and second the magnetic field has to be intense enough in order to induce a momentum asymmetry of the required magnitude.

We have made detailed calculations by means of the neutrino distribution function in the diffusion approximation, combined with the idea of a spherical resonance shell for neutrinos with momentum k that move in different directions relative to $\mathbf{B}$. This scheme provides a better description of the problem than the one formulated in terms of a single deformed sphere acting as an effective emission surface of sterile neutrinos. In our approach the sterile neutrinos produced in the resonance shell move in every direction relative to **B**. However, those going toward the interior of the protoneutron star cross again the resonance region and are reconverted into active neutrinos. These are within their own neutrinosphere and become thermalized. As a consequence, only the outgoing sterile neutrinos contribute to the total momentum emitted by the protostar.

Two opposite effects have to be taken into account when computing the fractional momentum asymmetry. One is purely geometric, and comes from the difference in the areas of the semispherical emission surface for neutrinos moving in opposite directions with the same momentum magnitude. The other is due to the radial temperature gradient, by which sterile neutrinos produced at different depths have unequal energy. In general both contributions do not compensate and thus we obtain a non null fractional asymmetry in the total momentum. Explicit results have been obtained for two simple models of a protostar, with the matter background assumed to be composed by nonrelativistic nucleons and degenerated electrons. One of them, which we worked with in detail, is the Eddington model for a spherical neutrino atmosphere.

Magnetic fields of the order of 10^{16}–10^{17} G are needed to reproduce the observed pulsar velocities. At first sight these fields could seem rather large, compared with the estimated values at the surface of a protoneutron star ($B_s \sim 10^{13}$ G), but in fact intensities as high as $B_c \sim 10^{18}$ G are possible at the core. A given parametrization for the magnetic field at the interior of a protoneutron star is[36]

$$B = B_s + B_c \left[1 - e^{-\beta(\rho/\rho_s)^\gamma}\right], \tag{98}$$

with $\beta \simeq 10^{-5}$ and $\gamma \simeq 3$. By adopting this profile as an upper boundary for the magnetic field in the protostar, we can see that the required field remains well below this boundary in most of the region within the neutrinosphere. In addition, the oscillation parameters are compatible with the allowed region for sterile neutrinos to be warm dark matter, leaving the mechanism as an attractive possibility to explain the proper motion of pulsars.

Acknowledgments

This work was partially supported by CONICET-Argentina, CONACYT- México, and DGAPA-UNAM (México) under grant PAPIIT-IN109001.

References

[1] A. Hewish, S. J. Bell, J. D. H. Pikington, P. F. Scott, and R. A. Collons, *Nature* **217**, 709 (1968).

[2] F. Pacini, *Nature* **216**, 567 (1966); T. Gold, *Nature* **218**, 731 (1966).

[3] S. L. Shapiro and S. A. Teukolsky, *Black Holes, White Dwarfs and Neutron Stars: The Physics of Compact Objects* (Wiley, 1983); N. K. Glendenning, Compact Stars: Nuclear Physics, Particle Physics, and General Relativity, 2nd Edition (Springer-Verlag, 2000).

[4] G. G. Raffelt, *Stars as Laboratories for Fundamental Physics* (The University of Chicago Press, Chicago, 1996); H. Suzuki, in *Physics and Astrophysics of Neutrinos*, edited by M. Fukugita and A. Suszuki (Springer-Verlag, 1994).

[5] A. G. Lyne and D. R. Lorimer, *Nature* **369**, 127 (1994); D. R. Lorimer, M. Bailes, and P. A. Harrison, *Mon. Not. R. Aston. Soc.* **289**, 592 (1997); B. M. S. Hansen and E. S. Phinney, *Mon. Not. R. Aston. Soc.* **291**, 569 (1997); J. M. Cordes and D. F. Chernoff, *Ap. J.* **505**, 315 (1998); C. Fryer, A. Burrows, and W. Benz, *Ap. J.* **496**, 333 (1998).

[6] Z. Arzoumanian, D. F. Chernoff, and J. M. Cordes, *Ap. J.* **568**, 289 (2002).

[7] See D. Lai, D. F. Chernoff, and J. M. Cordes, *Ap. J* **549**, 1111 (2001) and references therein.

[8] See, for example, D. Lai, *Neutron Star Kicks and Supernova Asymmetry*, to appear in the proceedings of 3-D Signatures in Stellar Explosions: A Workshop Honoring J. Craig Wheeler's 60th Birthday, Austin, Texas, 10-13 June 2003, astro-ph/0312542, and references therein.

[9] A. Burrows and J. Hayes, *Phys. Rev. Lett.* **76**, 352 (1996).

[10] M. Herant, W. Benz, W. R. Hix, C. L. Fryer., and S. A. Colgate, *Ap. J.* **435**, 339 (1994); A. Burrows, J. Hayes, B. A. Fryxell, *Ap. J.* **450**, 830 (1995); H-Th. Janka and E. Muller, *Astron. and Astrophys.* **290**, 496 (1994); *ibid.* **306**, 167 (1996); L. Scheck, T. Plewa, H.-Th. Janka, K. Kifonidis and E. Müller, *Phys. Rev. Lett.* **92**, 011103 (2004).

[11] C. L. Fryer, *Ap. J.* **601**, L175-L178 (2004).

[12] N. N. Chugai, *Sov. Astron. Lett.* **10**, 87 (1984); O. F. Dorofeev et al. *Sov. Astron. Lett.* **11**, 123 (1985); A. Vilenkin *Ap. J.* **451**, 700 (1995); C. J. Horowitz and G. Li, *Phys. Rev. Lett.* **80**, 3694 (1998); A. Kusenko, G. Segrè, and A. Vilenkin *Phys. Lett.* **B437**, 359 (1998); P. Arras and D. Lai, *Phys. Rev.* **D60**, 043001 (1999).

[13] G. S. Bisnovatyi-Kogan, *Astron. Astrophys. Trans.* **3**, 287 (1993); D. Lai and Y.-Z. Quian, *Ap. J.* **505**, 844 (1998); E. Roulet, *J. High Energy Phys.* **01**, 013 (1998).

[14] R. C. Duncan and C. Thompson, *Ap. J.* **392**, L9 (1992).

[15] V. B. Semikoz, *Sov. J. Nucl. Phys.* **46**, 946 (1987); J. C. D'Olivo, J. F. Nieves, and P. B. Pal, *Phys. Rev.* **D40**, 3679 (1989).

[16] J. C. D'Olivo and J. F. Nieves, *Phys. Rev.* **D56**, 5898 (1997).

[17] A. Kusenko and G. Segrè, *Phys. Rev. Lett.* **77**, 4872 (1996); ibid. 79, 2751 (1997); *Phys. Rev.* **D59**, 061302 (1999).

[18] Y. Z. Qian, *Phys. Rev. Lett.* **79**, 2750 (1997), C. W. Kim, J. D. Kim, and J. Song, *Phys. Lett.* **B419**, 279 (1998).

[19] M. Barkovich, J. C. D'Olivo, R. Montemayor, and J. F. Zanella, *Phys. Rev.* **D66**, 123005 (2002).

[20] A. Kusenko and G. Segrè, *Phys. Lett.* **B396**, 197 (1997).

[21] G. M. Fuller, A. Kusenko, I. Mocioiu, and S. Pascoli, *Phys. Rev.* **D68**, 103002 (2003); A. Kusenko, *Int. J. Mod. Phys.* **D13**, 2065 (2004).

[22] M. Barkovich, J. C. D'Olivo, and R. Montemayor, *Phys. Rev.* **D70**, 043005 (2004).

[23] For a recent review see M. Cirelli, astro-ph/0410122, to appear in the proceedings of 10th International Symposium on Particles, Strings and Cosmology (PASCOS 04 and Pran Nath Fest), Boston, Massachusetts, 16-22 August 2004.

[24] E. K. Akhmedov, A. Lanza, and D. W. Sciama, *Phys. Rev.* **D56**, 6117 (1997); E. Nardi and J. I. Zuluaga, *Ap. J.* **549**, 1076 (2001); G. Lambiase, astro-ph/0411242; G. Lambiase, astro-ph/0412408.

[25] D. Grasso, H. Nunokawa, and J. W. F. Valle, *Phys. Rev. Lett.* **81**, 2412 (1998).

[26] R. Horvat, *Mod. Phys. Lett.* **A13**, 2379 (1998); M. Barkovich, H. Casini, J. C. D'Olivo, and R. Montemayor, *Phys. Lett.* **B506**, 20 (2001).

[27] H. A. Bethe, *Rev. Mod. Phys* **62**, 801 (1990).

[28] P. Goldreich and S. V. Weber, *Astrophys. J.* **238**, 991 (1980).

[29] D. Z. Freedman, *Phys. Rev.* **D 9**, 1389 (1974).

[30] T. J. Mazurek, *Nature* **252**, 287 (1974); J. Cooperstein, *Phys. Rep.* **163**, 95 (1988).

[31] S. A. Colgate and M. H. Johnson, *Phys. Rev. Lett.* **5**, 235 (1960).

[32] A. Burrows and J. Lattimer, *Ap. J.* **307**, 178 (1986).

[33] A. Burrows and T. J. Mazurek, *Ap. J.* **259**, 330 (1982).

[34] P. Mészáros, *High-Energy Radiation from Magnetized Neutron Stars* (The University of Chicago Press, 1992).

[35] V. Urpin and J. Gil, *Astron. Astrophys.* **415**, 305 (2004).

[36] D. Bandyopadhyay, S. Chakrabarty and S. Pal, *Phys. Rev. Lett.* **79**, 2176 (1997). R. Manka, M. Zastawny-Kubika, A. Brzezina, I. Bednarek, astro-ph/0202303.

[37] B. Pontecorvo, *Zh. Eksp. Theor. Fiz.* **33**, 549 (1957); ibid.34, 247 (1958); Z. Maki, M. Nakagawa, and S. Sakata, *Prog. Theor. Phys.* **28**, 870 (1962).

[38] R. N. Mohapatra and P. Pal, *Massive Neutrinos in Physics and Astrophysics*, 3rd. Edition (World Scientific, 2004).

[39] Y. Hayato, The Super-Kamiokande and K2K Collaborations, *Eur. Phys. J.* **C33**, s829 (2004).

[40] A. B. Mc Donald, nucl-ex/0412005.

[41] D. Notzold and G. Raffelt, *Nucl. Phys.* **B307**, 924 (1988); J. C. D'Olivo, J. F. Nieves, and M. Torres, *Phys. Rev.* **D46**, 1172 (1992).

[42] L.Wolfenstein, *Phys. Rev* **D17**, 2369 (1978); S. P. Mikheyev y A. Yu. Smirnov, *Nuovo Cim.* **9C**, 17 (1986).

[43] A. Yu. Smirnov, hep-ph/0412391.

[44] J. C. D'Olivo and J. F. Nieves, *Phys. Lett.* **B383**, 87 (1996).

[45] E. Kh.Akhmedov, *Phys. Lett.* **B213**, 64 (1988); C. -S. Lim and W. J. Marciano, *Phys. Rev.* **D37**, 1368 (1988); M. M. Guzzo and H. Nonokawa, *Astrop. Phys.* **11**, 317 (1999); O. G. Miranda et al., *Nucl. Phys.* **B595**, 360 (2001).

[46] H. Nunokawa, V. B. Semikoz, A. Yu. Smirnov, and J. W. F. Valle, *Nucl. Phys.* **B501**, 17 (1997).

[47] V. B. Semikoz and J. W. F. Valle, *Nucl. Phys.* **B425**, 651 (1994); erratum *Nucl. Phys.* **B485**, 651 (1997); P. Elmfors, D. Grasso, and G. Raffelt, ibid. B479, 3 (1996); A. Erdas, C. W. Kim, and T. H. Lee, *Phys. Rev.* **D58**, 085016 (1998).

[48] P. J. Schinder and S. L. Shapiro, *Ap.J.* **259**, 311 (1982).

[49] A. D. Dolgov and S. H. Hansen, *Astroparticle Physics* **16**, 339 (2001).

In: Pulsars: Discoveries, Functions and Formation
Editor: Peter A. Travelle, pp. 97-123

ISBN 978-1-61122-982-0
© 2011 Nova Science Publishers, Inc.

Chapter 6

RADIATION OF THE GRAVITATIONAL AND ELECTROMAGNETIC BINARY PULSARS

Miroslav Pardy[*]
Institute of Plasma Physics ASCR, Prague Asterix Laser System,
PALS, Za Slovankou 3, 182 21 Prague 8, Czech Republic
and
Masaryk University, Department of, Physical Electronics,
Kotlářská 2, 611 37, Brno, Czech Republic

Abstract

The first part of this chapter gives the introductory words on pulsars. Then, in the second part of our text concerning the gravitational and electromagnetic radiation of pulsars, we have derived the total quantum energy loss of the binary. The energy loss is caused by the emission of gravitons during the motion of the two binary bodies around each other under their gravitational interaction. The energy-loss formula of the production of gravitons is derived here in the source theory formulation of gravity. Because the general relativity does not necessarily contain the last valid words to be written about the nature of gravity and it is not, of course, a quantum theory, it cannot give the answer on the production of gravitons and the quantum energy loss, respectively. So, this is the treatise that discusses the quantum energy loss caused by the production of gravitons by the binary system.

In the third part of this chapter, the quantum energy loss formula involving radiative corrections is derived also in the framework of the source theory. The general relativity does not necessarily contain the method how to express the quantum effects together with the radiative corrections by the geometrical language. So it cannot give the answer on the production of gravitons and on the graviton propagator with radiative corrections. So, the present text discusses the quantum energy loss caused by the production of gravitons and by the radiative corrections in the graviton propagator in case of the motion of binary.

At the present time the only classical radiation of gravity was confirmed. The production of gravitons is not involved in the Einstein theory. However, the idea, that radiative corrections can have macroscopic consequences is logically clear.

---

[*]E-mail address: pamir@physics.muni.cz, Phone: +420 54949 6393, Fax: +420541211214

We present an idea that radiative corrections effects might play a role in the physics of cosmological scale. The situation in the gravity problems with radiative corrections is similar to the situation in QED in time many years ago when the QED radiative corrections were theoretically predicted and then experimentally confirmed for instance in case of he Lamb shift, or, of the anomalous magnetic moment of electron.

Astrophysics is, therefore, in crucial position in proving the influence of radiative corrections on the dynamics in the cosmic space. We hope that the further astrophysical observations will confirm the quantum version of the energy loss of the binary with graviton propagator with radiative corrections.

In the fourth part of the chapter, we have derived the power spectrum formula of the synchrotron radiation generated by the electron and positron moving at the opposite angular velocities in homogeneous magnetic field. We have used the Schwinger version of quantum field theory, for its simplicity. It is surprising that the spectrum depends periodically on radiation frequency ω and time which means that the system composed from electron, positron and magnetic field behaves as a pulsating system. While such pulsar can be represented by a terrestrial experimental arrangement it is possible to consider also the cosmological analogue.

To our knowledge, our result is not involved in the classical monographs on the electromagnetic theory and at the same time it was not still studied by the accelerator experts investigating the synchrotron radiation of bunches. This effect was not described in textbooks on classical electromagnetic field and on the synchrotron radiation. We hope that sooner or later this effect will be verified by the accelerator physicists.

The treatise involves modification of an older article (Pardy, 1983a), in which the spectral formula of the gravitational radiation was derived, author's CERN preprints (Pardy, 1994a,b) and the author preprints and articles on the electromagnetic pulsars (Pardy, 2000a,b; 2001, 2002).

1.   Pulsars in General

Pulsars are specific cosmological objects which radiates electromagnetic energy in the form of short pulses. They were discovered by Hewish et all. in 1967, published by Hewish et al. (1968) and specified later as neutron stars. Now, it is supposed that they are fast rotating neutron stars with approximately Solar mass and with strong magnetic fields ($10^9 - 10^{14}$ Gauss). The neutron stars are formed during the evolution of stars and they are the product of the reaction $p + e^- \rightarrow n + \nu_e$, where the symbols in the last equation are as follows: proton, electron, neutron and the electron neutrino. The neutron stars were postulated by Landau (1932) at the year of the discovery of neutron in 1932. The pulsars composed from pions, or, hyperons, or, quarks probably also exist in universe, however it is not a measurement technique which rigorously determines these kinds of pulsars.

Pulsars emit highly accurate periodic signals mostly in radio waves beamed in a cone of radiation centered around their magnetic axis. These signals define the period of rotation of the neutron star, which radiates as a light-house once per revolution. We know so called slow rotated pulsars, or, normal pulsars with period P > 20ms and so called millisecond pulsars with period P < 20ms.

In 1974 a pulsar in a binary system was discovered by Hulse and Taylor and the discovery was published in 1975 (Hulse and Taylor, 1975). The period of rotation of normal pulsars increases with time and led to the rejection of the suggestion that the periodic signal could be due to the orbital period of binary stars. The orbital period of an isolated binary system decreases as it loses energy, whereas the period of a rotating body increases as it loses energy.

The present literature concerns only the electromagnetic pulsars. The number of them is 1300 and they were catalogued with very precise measurements of their positions and rotation rates. The published pulse profiles are so called integrated profiles obtained by adding some hundred of thousands of individual pulses. The integration hides a large variation of size and shape from pulse to pulse of the individual pulsar. The radiation is emitted along the direction of the field lines, so that the observed duration of the integrated profile depends on the inclination of the dipole axis to the rotation axis, because it is supposed that the pulsar radiation is a radiation of the dipole in the magnetic field. It means that the radiation of these pulsars is the synchrotron radiation of charged particles. We know that the analogical ultrarelativistic charge moving in a constant magnetic field radiates the synchrotron radiation in a very narrow cone and such system can be considered as a free electron laser in case that the opening angle of the cone is very small. Of course the angle of emission of pulsars is not smaller than radians. In other words the observed pulsars are not the free electron lasers. The idea that pulsar can be a cosmical maser was also rejected.

Pulsar radio emission is highly polarized, with linear and circular components. Individual pulses are often observed to be 100% polarized. The study of the polarization of pulsar is the starting point of the determination of their real structure.

The only energy source of the pulsar is the rotational energy of the neutron star. The rate of the dissipation of the rotational energy can be determined. The moment of inertia is fairly accurately known from the theory of the internal structure and the rotational slowdown is very accurately measured for almost every pulsar. So most of energy is radiated as magnetic dipole radiation at the rotation frequency leading to a measure of the magnetic dipole and the surface field strength.

The published articles on pulsars deal with the observation of the pulses and with the theoretical models. The observational results giving an insight into the behavior of matter in the presence of extreme gravitational and electromagnetic field are summarized for instance by Manchester (Manchester, 1992). The emission mechanism of photons are reviewed from a plasmatical viewpoint by Melrose (1992). The morfology of the radio pulsars is presented in the recent treatise by Seiradakis et. al. (2004). The review of properties of pulsars involving the radio propagation in the magnetosphere and of emission mechanism is summarized in the article by Graham-Smith (2003). At the same time there are, to our knowledge, no information on so called gravitational pulsars, or, on the models where the pulses are produced by the retrograde motion of the bodies moving around the central body.

So, the question arises, if it is possible to define gravitational binary pulsar, where the gravitational energy is generated by the binary system, or, by the system where two components are in the retrograde motion. We suppose that in case of the massive binary system the energy is generated in a cone starting from the component of a binary and it can be seen only if the observer is present in the axis this cone. Then, the observer detects gravitational pulses when the detector is sufficiently sensitive. There are many methods for the detection of the gravitational waves. One method, based on the quantum states of the superfluid ring, was suggested by author (Pardy, 1989).

We know that gravitational waves was indirectly confirmed by the observation of the period of the pulsar PSR 1913 + 16. The energy loss of this pulsar was calculated in the framework of the classical theory of gravitation. The quantum energy loss was given for instance by Manoukian (1990). His calculation was based on the so called Schwinger source theory where gravity is considered as a field theory of gravitons where graviton is a boson with spin 2, helicity ± 2 and zero mass. It is an analogue of photon in the electromagnetic theory.

In the following text we start with the source derivation of the power spectral formula of the gravitational radiation of a binary. Then, we calculate the quantum energy loss of a binary and the gravitational power spectrum involving radiative corrections. In the last part of a chapter, we consider so called electromagnetic pulsar which is formed by two particles with the opposite electrical charges which move in the constant magnetic field and generate the electromagnetic pulses.

2.   The Quantum Gravity Energy Loos of a Binary System

2.1.   Introduction

At the present time, the existence of gravitational waves is confirmed, thanks to the experimental proof of Taylor and Hulse who performed the systematic measurement of the motion of the binary with the pulsar PSR 1913+16. They found that the generalized energy-loss formula, which follows from the Einstein general theory of relativity, is in accordance with their measurement.

This success was conditioned by the fact that the binary with the pulsar PSR 1913+16 as a gigantic system of two neutron stars, emits sufficient gravitational radiation to influence the orbital motion of the binary at the observable scale.

Taylor and Hulse, working at the Arecibo radiotelescope, discovered the radiopulsar PSR 1913+16 in a binary, in 1974, and this is now considered as the best general relativistic laboratory (Taylor, 1993).

Pulsar PSR 1913+16 is the massive body of the binary system where each of the rotating pairs is 1.4 times the mass of the Sun. These neutron stars rotate around each other with a period 7.8 hours, in an orbit not much larger than Sun's diameter. Every 59 ms, the pulsar emits a short signal that is so clear that the arrival time of a 5-min string of a set of such signals can be resolved within 15 μs.

A pulsar model based on strongly magnetized, rapidly spinning neutron stars was soon established as consistent with most of the known facts (Huguenin et al., 1968); its electrodynamical properties were studied theoretically (Gold, 1968) and shown to be plausibly capable of generating broadband radio noise detectable over interstellar distances. The binary pulsar PSR 1913+16 is now recognized as the harbinger of a new class of unusually short-period pulsars, with numerous important applications.

Because the velocities and gravitational energies in a high-mass binary pulsar system can be significantly relativistic, strong-field and radiative effects come into play. The binary pulsar PSR 1913+16 provides significant tests of gravitation beyond the weak-field, slow-motion limit (Goldreich et al., 1969; Damour et al., 1992).

The goal of this section is not to repeat the derivation of the Einstein quadrupole formula, because this has been performed many times in general relativity and also in the Schwinger source theory in the weak-field limit (Manoukian, 1990).

We show that just in the framework of the source theory it is easy to determine the quantum energy-loss formula of the binary system. The energy-loss formula can be generalized in such a way it involves also the radiative corrections.

Since the measurement of the motion of the binaries goes on, we hope that future experiments will verify the quantum version of the energy-loss formula, involving also the radiative corrections.

2.2. The Source Theory Formulation of the Problem

We show how the total quantum loss of energy caused by the production of gravitons, emitted by the binary system of two point masses moving around each other under their gravitational interaction, can be calculated in the framework of the source theory of gravity.

Source theory (Schwinger, 1970, 1973, 1976) was initially constructed to describe the particle physics situations occurring in high-energy physics experiments. However, it was found that the original formulation simplifies the calculations in the electrodynamics and gravity, where the interactions are mediated by photon and graviton respectively. The source theory of gravity forms the analogue of quantum electrodynamics because, while in QED the interaction is mediated by the photon, the gravitational interaction is mediated by the graviton (Schwinger, 1976).

The source theory of gravity invented by Schwinger is linear theory. So, the question arises if it is in the coincidence with the Einstein gravity equations which are substantially nonlinear. The answer is affirmative, because the coincidence is only with the linear approximation of the Einstein theory. The experimental results of the Schwinger theory are also in harmony with experiment. The quadrupole formula of Einstein also follows from the Schwinger version.

The unification of gravity and electromagnetism is possible only in the Schwinger source theory. It is possible when, and only when it is possible the unification of forces. And this is performed in the Schwinger source theory of all interactions where force is of the Yukawa form. The problem of unification is not new. We know from the history of physics that the Ptolemy system could not be unified with the Galileo-Newton system because in the Ptolemy system it is not defined the force which is the fundamental quantity in the GN system and the primary cause of all phenomena in this system.

Einstein gravity uses the Riemann space-time where the gravity force has not the Yukawa dynamical form. The curvature of space-time is defined as the origin of all phenomena. The gravity force in the Einstein theory is in the antagonistic contradistinction with the Yukawa force in the quantum field theory and therefore it seems that QFT, QED, QCD and EL.-WEAK theory cannot be unified with the Einstein gravity.

Manoukian (1990) derivation of the Einstein quadrupole formula in the framework of the Schwinger source theory is possible because of the coincidence of the source theory with the linear limit of the Einstein theory.

Our approach is different from Manoukian method because we derive the power spectral spectral formula $P(\omega)$ of emitted gravitons with frequency ω and then using relation

$(-dE/dt) = \int d\omega P(\omega)$ we determine the energy loss E. In case of the radiative correction, we derive only the power spectral formula in the general form.

The mathematical structure of $P(\omega)$ follows directly from the action W and while in the case of the gravitational radiation, the formula is composed from the tensor of energy-momentum, then, in the case of the electromagnetic radiation formula, it involves charged vector currents.

The basic formula in the source theory is the vacuum-to-vacuum amplitude (Schwinger et al., 1976):

$$\langle 0_+|0_-\rangle = e^{\frac{i}{\hbar}W(S)}, \tag{1}$$

where the minus and plus tags on the vacuum symbol are causal labels, referring to any time before and after region of space-time, where sources are manipulated. The exponential form is introduced with regard to the existence of the physically independent experimental arrangements, which has the simple consequence that the associated probability amplitudes multiply and the corresponding W expressions add (Schwinger et al., 1976; Dittrich, 1978).

In the flat space-time, the field of gravitons is described by the amplitude (1) with the action (Schwinger, 1970) ($c = 1$ in the following text)

$$W(T) = 4\pi G \int (dx)(dx')$$

$$\times \left[T^{\mu\nu}(x)D_+(x-x')T_{\mu\nu}(x') - \frac{1}{2}T(x)D_+(x-x')T(x') \right], \tag{2}$$

where the dimensionality of $W(T)$ is the same as the dimensionality of the Planck constant $\hbar$; $T_{\mu\nu}$ is the tensor of momentum and energy. For a particle moving along the trajectory $\mathbf{x} = \mathbf{x}(t)$, it is defined by the equation (Weinberg, 1972):

$$T^{\mu\nu}(x) = \frac{p^\mu p^\nu}{E}\delta(\mathbf{x} - \mathbf{x}(t)), \tag{3}$$

where p^μ is the relativistic four-momentum of a particle with a rest mass m and

$$p^\mu = (E, \mathbf{p}) \tag{4}$$

$$p^\mu p_\mu = -m^2, \tag{5}$$

and the relativistic energy is defined by the known relation

$$E = \frac{m}{\sqrt{1 - \mathbf{v}^2}}, \tag{6}$$

where $\mathbf{v}$ is the three-velocity of the moving particle.

Symbol $T(x)$ in formula (2) is defined as $T = g_{\mu\nu}T^{\mu\nu}$, and $D_+(x - x')$ is the graviton propagator whose explicit form will be determined later.

The action W is not arbitrary because it must involve the attractive force between the gravity masses while in case of the electromagnetic situation the action must involve the repulsive force between charges of the same sign. It is very surprising that such form of

Lagrangians follows from the quantum definition of the vacuum to vacuum amplitude. It was shown by Schwinger that Einstein gravity also follows from the source theory, however the method of derivation is not the integral part of the source theory because the source theory is linear and it is not clear how to establish the equivalence between linear and nonlinear theory. String theory tries to solve the problem of the unification of all forces, however, this theory is, at the present time, not predictable and works with so called extra-dimensions which was not observed. It is not clear from the viewpoint of physics, what the dimension is. It seems that many problems can be solved in the framework of the source theory.

2.3. The Power Spectral Formula in General

It may be easy to show that the probability of the persistence of vacuum is given by the following formula (Schwinger et al., 1976):

$$|\langle 0_+|0_-\rangle|^2 = \exp\left\{-\frac{2}{\hbar}\mathrm{Im}W\right\} \stackrel{d}{=} \exp\left\{-\int dtd\omega \frac{1}{\hbar\omega}P(\omega,t)\right\}, \tag{7}$$

where the so-called power spectral function $P(\omega,t)$ has been introduced (Schwinger et al., 1976). In order to extract this spectral function from Im W, it is necessary to know the explicit form of the graviton propagator $D_+(x - x')$. The physical content of this propagator is analogous to the content of the photon propagator. It involves the gravitons property of spreading with velocity c. It means that its explicit form is just the same as that of the photon propagator. With regard to the source theory (Schwinger et al., 1976) the x-representation of $D_+(x)$ in eq. (2) is as follows:

$$D_+(x - x') = \int \frac{(dk)}{(2\pi)^4} e^{ik(x-x')} D(k), \tag{8}$$

where

$$D(k) = \frac{1}{|\mathbf{k}^2| - (k^0)^2 - i\epsilon}, \tag{9}$$

which gives

$$D_+(x - x') = \frac{i}{4\pi^2} \int_0^\infty d\omega \frac{\sin\omega|\mathbf{x} - \mathbf{x}'|}{|\mathbf{x} - \mathbf{x}'|} e^{-i\omega|t-t'|}. \tag{10}$$

Now, using formulas (2), (7) and (10), we get the power spectral formula in the following form:

$$P(\omega,t) = 4\pi G\omega \int (d\mathbf{x})(d\mathbf{x}')dt' \frac{\sin\omega|\mathbf{x} - \mathbf{x}'|}{|\mathbf{x} - \mathbf{x}'|} \cos\omega(t - t')$$

$$\times \left[T^{\mu\nu}(\mathbf{x},t)T_{\mu\nu}(\mathbf{x}',t') - \frac{1}{2}g_{\mu\nu}T^{\mu\nu}(\mathbf{x},t)g_{\alpha\beta}T^{\alpha\beta}(\mathbf{x}',t')\right]. \tag{11}$$

2.4. The Power Spectral Formula for the Binary System

In the case of the binary system with masses m_1 and m_2, we suppose that they move in a uniform circular motion around their center of gravity in the xy plane, with corresponding kinematical coordinates:

$$\mathbf{x}_1(t) = r_1(\mathbf{i}\cos(\omega_0 t) + \mathbf{j}\sin(\omega_0 t)) \tag{12}$$

$$\mathbf{x}_2(t) = r_2(\mathbf{i}\cos(\omega_0 t + \pi) + \mathbf{j}\sin(\omega_0 t + \pi)) \tag{13}$$

with

$$\mathbf{v}_i(t) = d\mathbf{x}_i/dt, \quad \omega_0 = v_i/r_i, \quad v_i = |\mathbf{v}_i| \quad (i = 1, 2). \tag{14}$$

For the tensor of energy and momentum of the binary we have:

$$T^{\mu\nu}(x) = \frac{p_1^\mu p_1^\nu}{E_1}\delta(\mathbf{x} - \mathbf{x}_1(t)) + \frac{p_2^\mu p_2^\nu}{E_2}\delta(\mathbf{x} - \mathbf{x}_2(t)), \tag{15}$$

where we have omitted the tensor $t_{\mu\nu}^G$, which is associated with the massless, gravitational field distributed all over space and proportional to the gravitational constant G (Cho et al., 1976):

After insertion of eq. (15) into eq. (11), we get:

$$P_{total}(\omega, t) = P_1(\omega, t) + P_{12}(\omega, t) + P_2(\omega, t), \tag{16}$$

where ($t' - t = \tau$):

$$P_1(\omega, t) = \frac{G\omega}{r_1\pi}\int_{-\infty}^\infty d\tau \frac{\sin[2\omega r_1\sin(\omega_0\tau/2)]}{\sin(\omega_0\tau/2)}\cos\omega\tau$$

$$\times \left(E_1^2(\omega_0^2 r_1^2\cos\omega_0\tau - 1)^2 - \frac{m_1^4}{2E_1^2}\right), \tag{17}$$

$$P_2(\omega, t) = \frac{G\omega}{r_2\pi}\int_{-\infty}^\infty d\tau \frac{\sin[2\omega r_2\sin(\omega_0\tau/2)]}{\sin(\omega_0\tau/2)}\cos\omega\tau$$

$$\times \left(E_2^2(\omega_0^2 r_2^2\cos\omega_0\tau - 1)^2 - \frac{m_2^4}{2E_2^2}\right), \tag{18}$$

$$P_{12}(\omega, t) = \frac{4G\omega}{\pi}\int_{-\infty}^\infty d\tau \frac{\sin\omega[r_1^2 + r_2^2 + 2r_1r_2\cos(\omega_0\tau)]^{1/2}}{[r_1^2 + r_2^2 + 2r_1r_2\cos(\omega_0\tau)]^{1/2}}\cos\omega\tau$$

$$\times \left(E_1E_2(\omega_0^2 r_1r_2\cos\omega_0\tau + 1)^2 - \frac{m_1^2 m_2^2}{2E_1E_2}\right). \tag{19}$$

2.5.   The Quantum Energy Loss of the Binary

Using the following relations

$$\omega_0 \tau = \varphi + 2\pi l, \qquad \varphi \in (-\pi, \pi), \quad l = 0, \pm 1, \pm 2, \ldots \tag{20}$$

$$\sum_{l=-\infty}^{l=\infty} \cos 2\pi l \frac{\omega}{\omega_0} = \sum_{l=-\infty}^{\infty} \omega_0 \delta(\omega - \omega_0 l), \tag{21}$$

we get for $P_i(\omega, t)$, with ω being restricted to positive:

$$P_i(\omega, t) = \sum_{l=1}^{\infty} \delta(\omega - \omega_0 l) P_{il}(\omega, t). \tag{22}$$

Using the definition of the Bessel function $J_{2l}(z)$

$$J_{2l}(z) = \frac{1}{2\pi} \int_{-\pi}^{\pi} d\varphi \cos\left(z \sin \frac{\varphi}{2}\right) \cos l\varphi, \tag{23}$$

from which the derivatives and their integrals follow, we get for P_{1l} and P_{2l} the following formulas:

$$P_{il} = \frac{2G\omega}{r_i}\left(\left(E_i^2(1 - v_i^2)^2 - \frac{m_i^4}{2E_i^2}\right)\int_0^{2v_i l} dx\, J_{2l}(x)\right.$$

$$+ \quad 4E_i^2(1 - v_i^2)v_i^2 J'_{2l}(2v_i l) + 4E_i^2 v_i^4 J'''_{2l}(2v_i l)\bigg), \quad i = 1, 2. \tag{24}$$

Using $r_2 = r_1 + \epsilon$, where ϵ is supposed to be small in comparison with radii r_1 and r_2, we obtain

$$[r_1^2 + r_2^2 + 2r_1 r_2 \cos \varphi]^{1/2} \approx 2a \cos\left(\frac{\varphi}{2}\right), \tag{25}$$

with

$$a = r_1\left(1 + \frac{\epsilon}{2r_1}\right). \tag{26}$$

So, instead of eq. (19) we get:

$$P_{12}(\omega, t) = \frac{2G\omega}{a\pi} \int_{-\infty}^{\infty} d\tau \frac{\sin[2\omega a \cos(\omega_0 \tau/2)]}{\cos(\omega_0 \tau)/2]} \cos \omega\tau$$

$$\times \quad \left(E_1 E_2(\omega_0^2 r_1 r_2 \cos \omega_0 \tau + 1)^2 - \frac{m_1^2 m_2^2}{2E_1 E_2}\right). \tag{27}$$

Now, we can approach the evaluation of the energy-loss formula for the binary from the power spectral formulas (24) and (27). The energy loss is defined by the relation

$$-\frac{dU}{dt} = \int P(\omega)d\omega =$$

$$\int d\omega \sum_{i,l} \delta(\omega - \omega_0 l) P_{il} + \int P_{12}(\omega) d\omega = -\frac{d}{dt}(U_1 + U_2 + U_{12}). \qquad (28)$$

Or,

$$-\frac{d}{dt}U_i = \int d\omega \sum_l \delta(\omega - \omega_0 l) P_{il}, \quad -\frac{d}{dt}U_{12} = \int d\omega \sum_{i,l} \delta(\omega - \omega_0 l) P_{12l}. \qquad (29)$$

From Sokolov and Ternov (1983) we learn Kapteyn's formulas:

$$\sum_{l=1}^{\infty} 2l J_{2l}'(2lv) = \frac{v}{(1 - v^2)^2}, \qquad (30)$$

and

$$\sum_{l=1}^{\infty} l \int_0^{2lv} J_{2l}(x) dx = \frac{v^3}{3(1 - v^2)^3}. \qquad (31)$$

The formula $\sum_{l=1}^{\infty} l J_{2l}'''(2lv) = 0$ can be obtained from formula

$$\sum_{l=1}^{\infty} \frac{1}{l} J_{2l}'(2lv) = \frac{1}{2}v^2 \qquad (32)$$

by its differentiation with the respect to v (Schott, 1912).

Then, after application of eqs. (30), (31) and (32) to eqs. (24) and (28), we get:

$$-\frac{dU_i}{dt} = \frac{2G\omega_0}{r_i} \left[\left(E_i^2(v_i^2 - 1)^2 - \frac{m_i^4}{2E_i^2} \right) \frac{v_i^3}{3(1 - v_i^2)^3} - 2E_i^2 v_i^3 + 4E_i^2 v_i^4 \right]. \qquad (33)$$

Instead of using Kapteyn's formulas for the interference term, we will perform a direct evaluation of the energy loss of the interference term by the ω-integration in (27). So, after some elementary modification in the ω-integral, we get:

$$-\frac{dU_{12}}{dt} = \int_0^{\infty} P(\omega) d\omega =$$

$$A \int_{-\infty}^{\infty} d\tau \int_{-\infty}^{\infty} d\omega \omega e^{-i\omega\tau} \sin[2\omega a \cos\omega_0\tau] \left[\frac{B(C \cos\omega_0\tau + 1)^2 - D}{\cos(\omega_0\tau/2)} \right], \qquad (34)$$

with

$$A = \frac{G}{a\pi}, \quad B = E_1 E_2, \quad C = v_1 v_2, \quad D = \frac{m_1^2 m_2^2}{2E_1 E_2}. \qquad (35)$$

Using the definition of the δ-function and its derivative, we have, instead of eq. (34), with $v = a\omega_0$:

$$-\frac{dU_{12}}{dt} = A\omega_0\pi \int_{-\infty}^{\infty} dx \, \frac{[B(C\cos x + 1)^2 - D]}{\cos(x/2)} \quad \times$$

$$[\delta'(x - 2v\cos(x/2)) - \delta'(x + 2v\cos(x/2))]. \tag{36}$$

Putting

$$x - 2v\cos(x/2) = t, \tag{37}$$

in the first δ'-term and

$$y + 2v\cos(y/2) = t, \tag{38}$$

in the second δ'-term, we get eq. (36) in the following form:

$$-\frac{dU_{12}}{dt} = 2A\omega_0 v\pi \int_{-\infty}^{\infty} dt\delta'(t) \times$$

$$\left\{\frac{[B(C\cos x + 1)^2 - D]}{(x - t)(1 + v\sin(x/2))} - \frac{[B(C\cos y + 1)^2 - D]}{(y + t)(1 - v\sin(y/2))}.\right\} \tag{39}$$

Using the known relation for a δ-function:

$$\int dt f(t)\delta'(t) = -f'(0), \tag{40}$$

we get the energy loss formula for the synergic term in the form:

$$-\frac{dU_{12}}{dt} = -2A\omega_0 v\pi \frac{d}{dt}\left\{\frac{[B(C\cos x + 1)^2 - D]}{(x - t)(1 + v\sin(x/2))} - \frac{[B(C\cos y + 1)^2 - D]}{(y - t)(1 - v\sin(y/2))}\right\}\bigg|_{t=0}, \tag{41}$$

and we recommend the final calculation of the last formula to the mathematical experts.

Let us remark finally that the formulas derived for the energy loss of the binary (33) and (41) describes only the binary system and therefore their sum has not the form of the Einstein quadrupole formula. The sum forms the total produced gravitational energy, and involves not only the radiation of the individual bodies of the binary, but also the interference term. The problem of the coincidence with the Einstein quadrupole formula is open.

3.   The Power Spectal Formula Ivolving Radiative Corrections

3.1.   Introduction

We here calculate the total quantum loss of energy caused by production of gravitons emitted by the binary system in the framework of the source theory of gravity for the situation with the gravitational propagator involving radiative corrections.

We know from QED that photon can exist in the virtual state as the two body system in the form of the electron positron pair. It means that the photon propagator involves the additional process:

$$\gamma \to e^+ + e^- \to \gamma. \tag{42}$$

In case of the graviton radiation, the situation is analogical with the situation in QED. Instead of eq. (42) we write

$$g \to 2e^+ + 2e^- \to g, \tag{43}$$

where g is graviton and number 2 is there in order to conserve spin also during the virtual process.

Equation (43) can be of course expressed in more detail:

$$g \to \gamma + \gamma \to (e^+ + e^-) + (e^+ + e^-) \to \gamma + \gamma \to g. \tag{44}$$

We will show that in the framework of the source theory it is easy to determine the quantum energy loss formula of the binary system both in case with the graviton propagator with radiative corrections.

We will investigate how the spectrum of the gravitational radiation is modified if we involve radiation corrections corresponding to the virtual pair production and annihilation in the graviton propagator. Our calculation is an analogue of the photon propagator with radiative corrections for production of photons by the Čerenkov mechanism (Pardy, 1994c,d).

Because the measurement of motion of the binaries goes on, we hope that the future experiments will verify the quantum version of the energy loss formula following from the source theory and that sooner or later the confirmation of this formula will be established.

3.2.   The Binary Power Spectrum with Radiative Corrections

According to source theory (Schwinger, 1973; Dittrich, 1978; Pardy, 1994c,d), the photon propagator in the Minkowski space-time with radiative correction is in the momentum representation of the form:

$$\tilde{D}(k) = D(k) + \delta D(k), \tag{45}$$

or,

$$\tilde{D}(k) = \frac{1}{|\mathbf{k}|^2 - (k^0)^2 - i\epsilon}$$

$$+ \int_{4m^2}^{\infty} dM^2 \frac{a(M^2)}{|\mathbf{k}|^2 - (k^0)^2 + \frac{M^2 c^2}{\hbar^2} - i\epsilon}, \tag{46}$$

where m is mass of electron and the last term in equation (44) is derived on the virtual photon condition

$$|\mathbf{k}|^2 - (k^0)^2 = -\frac{M^2 c^2}{\hbar^2}. \tag{47}$$

The weight function $a(M^2)$ has been derived in the following form (Schwinger, 1973; Dittrich, 1976):

$$a(M^2) = \frac{\alpha}{3\pi} \frac{1}{M^2} \left(1 + \frac{2m^2}{M^2}\right) \left(1 - \frac{4m^2}{M^2}\right)^{1/2}. \tag{48}$$

We suppose that the graviton propagator with the radiative corrections forms the analogue of the photon propagator.

Now, with regard to the definition of the Fourier transform

$$D_+(x - x') = \int \frac{(dk)}{(2\pi)^4} e^{ik(x-x')} D(k), \tag{49}$$

we get for δD_+ the following relation ($c = \hbar = 1$):

$$\delta D_+(x - x') = \frac{i}{4\pi^2} \int_{4m^2}^{\infty} dM^2 a(M^2)$$

$$\times \int d\omega \, \frac{\sin\left\{[\omega^2 - M^2]^{1/2}|\mathbf{x} - \mathbf{x}'|\right\}}{|\mathbf{x} - \mathbf{x}'|} e^{-i\omega|t-t'|}. \tag{50}$$

The function (50) differs from the gravitational function "D_+" in (9) especially by the factor

$$\left(\omega^2 - M^2\right)^{1/2} \tag{51}$$

in the function 'sin' and by the additional mass-integral which involves the radiative corrections to the original power spectrum formula.

In order to determine the additional spectral function of produced gravitons, corresponding to the radiative corrections, we insert $D_+(x - x') + \delta D_+(x - x')$ into eq. (2), and using eq. (11) we obtain (factor 2 from the two photons is involved):

$$\delta P(\omega, t) = \frac{4G\omega}{\pi} \int (d\mathbf{x})(d\mathbf{x}') dt' \int_{4m^2}^{\infty} dM^2 a(M^2)$$

$$\times \frac{\sin\left\{[\omega^2 - M^2]^{1/2}|\mathbf{x} - \mathbf{x}'|\right\}}{|\mathbf{x} - \mathbf{x}'|} \cos\omega(t - t')$$

$$\times \left[T^{\mu\nu}(\mathbf{x}, t)g_{\mu\alpha}g_{\nu\beta}T^{\alpha\beta}(\mathbf{x}', t') - \frac{1}{2}g_{\mu\nu}T^{\mu\nu}(\mathbf{x}, t)g_{\alpha\beta}T^{\alpha\beta}(\mathbf{x}', t')\right]. \tag{52}$$

Then using eqs. (16), (17), (18) and (19), we get

$$\delta P_{total}(\omega, t) = \delta P_1(\omega, t) + \delta P_2(\omega, t) + \delta P_{12}(\omega, t), \tag{53}$$

where ($t' - t = \tau$):

$$\delta P_1(\omega, t) = \frac{2G\omega}{r_1\pi} \int_{-\infty}^{\infty} d\tau \int_{4m^2}^{\infty} dM^2 a(M^2) \frac{\sin\{2\left(\omega^2 - M^2\right)^{1/2} r_1 \sin(\omega_0\tau/2)\}}{\sin(\omega_0\tau/2)} \cos\omega\tau$$

$$\times \left(E_1^2(\omega_0^2 r_1^2 \cos\omega_0\tau - 1)^2 - \frac{m_1^4}{2E_1^2}\right), \tag{54}$$

$$\delta P_2(\omega,t) = \frac{2G\omega}{r_2\pi} \int_{-\infty}^{\infty} d\tau \int_{4m^2}^{\infty} dM^2 a(M^2) \frac{\sin\{2\left(\omega^2 - M^2\right)^{1/2} r_2 \sin(\omega_0\tau/2)\}}{\sin(\omega_0\tau/2)} \cos \omega\tau$$

$$\times \left(E_2^2(\omega_0^2 r_2^2 \cos \omega_0\tau - 1)^2 - \frac{m_2^4}{2E_2^2} \right), \tag{55}$$

$$\delta P_{12}(\omega,t) = \frac{8G\omega}{\pi} \int_{-\infty}^{\infty} d\tau \int_{4m^2}^{\infty} dM^2 a(M^2)$$

$$\frac{\sin\{\left(\omega^2 - M^2\right)^{1/2} [r_1^2 + r_2^2 + 2r_1r_2 \cos(\omega_0\tau)]^{1/2}\}}{[r_1^2 + r_2^2 + 2r_1r_2 \cos(\omega_0\tau)]^{1/2}} \cos \omega\tau$$

$$\times \left(E_1 E_2(\omega_0^2 r_1 r_2 \cos \omega_0\tau + 1)^2 - \frac{m_1^2 m_2^2}{2E_1 E_2} \right). \tag{56}$$

The explicit determination of the power spectrum is the problem which was solved by author in 1994. The solution was performed only approximately. Here also can be expected only the approximative solution. From this solution can be then derived the energy loss as in the previous chapter.

Let us show the possible way of the determination of the spectral formula. If we introduce the new variable s by the relation

$$\omega^2 - M^2 = s^2; \quad -dM^2 = 2s ds \tag{57}$$

then, instead of eqs. (54), (55) and (56) we have

$$\delta P_1(\omega,t) = \frac{2G\omega}{r_1\pi} \int_{-\infty}^{\infty} d\tau \int_{s_1}^{s_2} (2s ds) a(\omega^2 - s^2) \frac{\sin\{2sr_1 \sin(\omega_0\tau/2)\}}{\sin(\omega_0\tau/2)} \cos \omega\tau$$

$$\times \left(E_1^2(\omega_0^2 r_1^2 \cos \omega_0\tau - 1)^2 - \frac{m_1^4}{2E_1^2} \right), \tag{58}$$

$$\delta P_2(\omega,t) = \frac{2G\omega}{r_2\pi} \int_{-\infty}^{\infty} d\tau \int_{s_1}^{s_2} (2s ds) a(\omega^2 - s^2) \frac{\sin\{2sr_2 \sin(\omega_0\tau/2)\}}{\sin(\omega_0\tau/2)} \cos \omega\tau$$

$$\times \left(E_2^2(\omega_0^2 r_2^2 \cos \omega_0\tau - 1)^2 - \frac{m_2^4}{2E_2^2} \right), \tag{59}$$

$$\delta P_{12}(\omega,t) = \frac{8G\omega}{\pi} \int_{-\infty}^{\infty} d\tau \int_{s_1}^{s_2} (2s ds) a(\omega^2 - s^2) \frac{\sin\{2s[r_1^2 + r_2^2 + 2r_1r_2 \cos(\omega_0\tau)]^{1/2}\}}{[r_1^2 + r_2^2 + 2r_1r_2 \cos(\omega_0\tau)]^{1/2}}$$

$$\times \cos \omega\tau \left(E_1 E_2(\omega_0^2 r_1 r_2 \cos \omega_0\tau + 1)^2 - \frac{m_1^2 m_2^2}{2E_1 E_2} \right), \tag{60}$$

where

$$s_1 = \omega^2 - 4m^2, \quad s_2 = \infty. \tag{61}$$

It seems that the rigorous procedure is the τ-integration as the first step and then s-integration as the second step (Pardy, 1994c,d). While, in case of the linear motion the mathematical operations are easy (Pardy, 1994c), in case of the circular motion there are some difficulties (Pardy, 1994d). The final form of eq. (60) is recommended for the mathematical experts.

The energy loss is given as follows:

$$-\delta\frac{dU_i}{dt} = \int_0^\infty d\omega\, \delta P_i(\omega, t); \quad -\delta\frac{dU_{12}}{dt} = \int_0^\infty d\omega\, \delta P_{12}(\omega, t). \tag{62}$$

Now, let us go to the discussion on the electromagnetic system of the two opposite charges moving in the constant magnetic field and producing the pulse synchrotron radiation.

4.   Electromagnetic Pulsar

4.1.   Introduction

Here, the power spectrum formula of the synchrotron radiation generated by the electron and positron moving at the opposite angular velocities in homogeneous magnetic field is derived in the Schwinger version of quantum field theory.

It is surprising that the spectrum depends periodically on radiation frequency ω and time which means that the system composed from electron, positron and magnetic field forms the pulsar.

We will show that the large hadron collider (LHC) which is at present time under construction in CERN can be considered in near future also as the largest electromagnetic terrestrial pulsar. We know that while the Fermilab's Tevatron handles counter-rotating protons and antiprotons in a single beam channel, LHC will operate with proton and proton beams in such a way that the collision center of mass energy will be 14 TeV and luminosity $10^{34}\mathrm{cm}^{-1}\mathrm{s}^{-2}$. To achieve such large luminosity it must operate with more than 2800 bunches per beam and a very high density of particles in bunches. The LHC will also operate for heavy Pb ion physics at a luminosity of $10^{27}\mathrm{cm}^{-1}\mathrm{s}^{-2}$ (Evans, 1999).

The collision of particles is caused by the opposite directional motion of bunches. Or, if one bunch has the angular velocity ω, then the bunch with antiparticles has angular velocity $-\omega$. Here we will determine the spectral density of emitted photons in the simplified case where one electron and one positron move in the opposite direction on a circle. We will show that the synergic spectrum depends periodically on time. This means that the behavior of the system is similar to the behavior of electromagnetic pulsar. The derived spectral formula describes the spectrum of photons generated by the Fermilab Tevatron. In case that the particles in bunches are of the same charge as in LHC, then, it is necessary to replace the function sine by cosine in the final spectral formula. Now let us approach the theory and explicit calculation of the spectrum.

This process is the generalization of the one-charge synergic synchrotron-Čerenkov radiation which has been calculated in source theory two decades ago by Schwinger et al. (1976). We will follow the Schwinger article and also the author articles (Pardy, 1994d, 2000, 2002) as the starting point. Although our final problem is the radiation of the two-charge system in vacuum, we consider, first in general, the presence of dielectric medium, which is represented by the phenomenological index of refraction n and it is well known that this phenomenological constant depends on the external magnetic field. Introducing the phenomenological constant enables to consider also the Čerenkovian processes. Later we put $n = 1$.

We will investigate here how the original Schwinger (et al.) spectral formula of the synergic synchrotron Čerenkov radiation of the charged particle is modified if we consider the electron and positron moving at the opposite angular velocities. This problem is an analogue of the linear (Pardy, 1997) and circular problem solved by author (Pardy, 2000). We will show that the original spectral formula of the synergic synchrotron-Čerenkov radiation is modulated by function $4\sin^2(\omega t)$ where ω is the frequency of the synergistic radiation produced by the system and it does not depend on the orbital angular frequency of electron or positron. We will use here the fundamental ingredients of Schwinger source theory (Schwinger, 1970, 1973; Dittrich, 1978; Pardy, 1994c, d, e) to determine the power spectral formula.

4.2. Formulation of the Electromagnetic Problem

The basic formula of the Schwinger source theory is the so called vacuum to vacuum amplitude: $\langle 0_+|0_-\rangle = \exp\{\frac{i}{\hbar}W\}$, where in case of the electromagnetic field in the medium, the action W is given by the following formula:

$$W = \frac{1}{2c^2}\int (dx)(dx')J^\mu(x)D_{+\mu\nu}(x-x')J^\nu(x'),\qquad(63)$$

where

$$D_+^{\mu\nu} = \frac{\mu}{c}[g^{\mu\nu} + (1-n^{-2})\beta^\mu\beta^\nu]D_+(x-x'),\qquad(64)$$

where $\beta^\mu \equiv (1,0)$, $J^\mu \equiv (c\varrho, \mathbf{J})$ is the conserved current, μ is the magnetic permeability of the medium, ϵ is the dielectric constant od the medium and $n = \sqrt{\epsilon\mu}$ is the index of refraction of the medium. Function D_+ is defined as in eq. (10) (Schwinger et al., 1976):

$$D_+(x-x') = \frac{i}{4\pi^2 c}\int_0^\infty d\omega \frac{\sin\frac{n\omega}{c}|\mathbf{x}-\mathbf{x}'|}{|\mathbf{x}-\mathbf{x}'|}e^{-i\omega|t-t'|}.\qquad(65)$$

The probability of the persistence of vacuum follows from the vacuum amplitude (1) where Im W is the basis for the following definition of the spectral function $P(\omega, t)$:

$$-\frac{2}{\hbar}\text{Im }W \stackrel{d}{=} -\int dt d\omega \frac{P(\omega, t)}{\hbar\omega}.\qquad(66)$$

Now, if we insert eq. (64) into eq. (66), we get after extracting $P(\omega, t)$ the following general expression for this spectral function:

$$P(\omega, t) = -\frac{\omega}{4\pi^2}\frac{\mu}{n^2}\int dx dx' dt' \left[\frac{\sin\frac{n\omega}{c}|\mathbf{x} - \mathbf{x}'|}{|\mathbf{x} - \mathbf{x}'|}\right]$$

$$\times \quad \cos[\omega(t - t')][\varrho(\mathbf{x}, t)\varrho(\mathbf{x}', t') - \frac{n^2}{c^2}\mathbf{J}(\mathbf{x}, t) \cdot \mathbf{J}(\mathbf{x}', t')], \tag{67}$$

which is an analogue of the formula (11).

Let us recall that the last formula can be derived also in the classical electrodynamic context as it is shown for instance in the Schwinger article (Schwinger, 1949). The derivation of the power spectral formula from the vacuum amplitude is more simple.

4.3.   The Radiation of Two Opposite Charges

Now, we will apply the formula (67) to the two-body system with the opposite charges moving at the opposite angular velocities in order to get in general synergic synchrotron-Čerenkov radiation of electron and positron moving in a uniform magnetic field

While the synchrotron radiation is generated in a vacuum, the synergic synchrotron-Čerenkov radiation can produced only in a medium with dielectric constant n. We suppose the circular motion with velocity $\mathbf{v}$ in the plane perpendicular to the direction of the constant magnetic field $\mathbf{H}$ (chosen to be in the $+z$ direction).

We can write the following formulas for the charge density ϱ and for the current density $\mathbf{J}$ of the two-body system with opposite charges and opposite angular velocities:

$$\varrho(\mathbf{x}, t) = e\delta(\mathbf{x} - \mathbf{x_1}(t)) - e\delta(\mathbf{x} - \mathbf{x_2}(t)) \tag{68}$$

and

$$\mathbf{J}(\mathbf{x}, t) = e\mathbf{v}_1(t)\delta(\mathbf{x} - \mathbf{x_1}(t)) - e\mathbf{v}_2(t)\delta(\mathbf{x} - \mathbf{x_2}(t)) \tag{69}$$

with

$$\mathbf{x}_1(t) = \mathbf{x}(t) = R(\mathbf{i}\cos(\omega_0 t) + \mathbf{j}\sin(\omega_0 t)), \tag{70}$$

$$\mathbf{x}_2(t) = R(\mathbf{i}\cos(-\omega_0 t) + \mathbf{j}\sin(-\omega_0 t)) = \mathbf{x}(-\omega_0, t) = \mathbf{x}(-t). \tag{71}$$

The absolute values of velocities of both particles are the same, or $|\mathbf{v}_1(t)| = |\mathbf{v}_2(t)| = v$, where ($H = |\mathbf{H}|$, E = energy of a particle)

$$\mathbf{v}(t) = d\mathbf{x}/dt, \quad \omega_0 = v/R, \quad R = \frac{\beta E}{eH}, \quad \beta = v/c, \quad v = |\mathbf{v}|. \tag{72}$$

After insertion of eqs. (68)–(71) into eq. (67), and after some mathematical operations we get

$$P(\omega, t) = -\frac{\omega}{4\pi^2}\frac{\mu}{n^2}e^2\int_{-\infty}^{\infty} dt' \cos(t - t') \sum_{i,j=1}^{2} (-1)^{i+j}$$

$$\times \quad \left[1 - \frac{\mathbf{v}_i(t) \cdot \mathbf{v}_j(t')}{c^2} n^2\right] \left\{\frac{\sin \frac{n\omega}{c}|\mathbf{x}_i(t) - \mathbf{x}_j(t')|}{|\mathbf{x}_i(t) - \mathbf{x}_j(t')|}\right\}. \tag{73}$$

Let us remark, that for situation of the identical charges, the factor $(-1)^{i+j}$ must be replaced by 1.

Using $t' = t + \tau$, we get for

$$\mathbf{x}_i(t) - \mathbf{x}_j(t') \overset{d}{=} \mathbf{A}_{ij}, \tag{74}$$

$$|\mathbf{A}_{ij}| = [R^2 + R^2 - 2RR\cos(\omega_0\tau + \alpha_{ij})]^{1/2} = 2R\left|\sin\left(\frac{\omega_0\tau + \alpha_{ij}}{2}\right)\right|, \tag{75}$$

where α_{ij} were evaluated as follows:

$$\alpha_{11} = 0, \quad \alpha_{12} = 2\omega_0 t, \quad \alpha_{21} = 2\omega_0 t, \quad \alpha_{22} = 0. \tag{76}$$

Using

$$\mathbf{v}_i(t) \cdot \mathbf{v}_j(t + \tau) = \omega_0^2 R^2 \cos(\omega_0\tau + \alpha_{ij}), \tag{77}$$

and relation (75) we get with $v = \omega_0 R$

$$P(\omega, t) = -\frac{\omega}{4\pi^2} \frac{\mu}{n^2} e^2 \int_{-\infty}^{\infty} d\tau \cos\omega\tau \sum_{i,j=1}^{2} (-1)^{i+j}$$

$$\times \quad \left[1 - \frac{n^2}{c^2} v^2 \cos(\omega_0\tau + \alpha_{ij})\right] \left\{\frac{\sin\left[\frac{2Rn\omega}{c}\sin\left(\frac{(\omega_0\tau + \alpha_{ij})}{2}\right)\right]}{2R\sin\left(\frac{(\omega_0\tau + \alpha_{ij})}{2}\right)}\right\}. \tag{78}$$

Introducing new variable T by relation

$$\omega_0\tau + \alpha_{ij} = \omega_0 T \tag{79}$$

for every integral in eq. (78), we get $P(\omega, t)$ in the following form

$$P(\omega, t) = -\frac{\omega}{4\pi^2} \frac{e^2}{2R} \frac{\mu}{n^2} \int_{-\infty}^{\infty} dT \sum_{i,j=1}^{2} (-1)^{i+j}$$

$$\times \quad \cos(\omega T - \frac{\omega}{\omega_0}\alpha_{ij})\left[1 - \frac{c^2}{n^2}v^2 \cos(\omega_0 T)\right] \left\{\frac{\sin\left[\frac{2Rn\omega}{c}\sin\left(\frac{\omega_0 T}{2}\right)\right]}{\sin\left(\frac{\omega_0 T}{2}\right)}\right\}. \tag{80}$$

The last formula can be written in the more compact form,

$$P(\omega, t) = -\frac{\omega}{4\pi^2} \frac{\mu}{n^2} \frac{e^2}{2R} \sum_{i,j=1}^{2} (-1)^{i+j} \left\{P_1^{(ij)} - \frac{n^2}{c^2}v^2 P_2^{(ij)}\right\}, \tag{81}$$

where

$$P^{(ij)} = J_{1a}^{(ij)} \cos \frac{\omega}{\omega_0} \alpha_{ij} + J_{1b}^{(ij)} \sin \frac{\omega}{\omega_0} \alpha_{ij} \tag{82}$$

and

$$P_2^{(ij)} = J_{2A}^{(ij)} \cos \frac{\omega}{\omega_0} \alpha_{ij} + J_{2B}^{(ij)} \sin \frac{\omega}{\omega_0} \alpha_{ij}, \tag{83}$$

where

$$J_{1a}^{(ij)} = \int_{-\infty}^{\infty} dT \cos \omega T \left\{ \frac{\sin \left[\frac{2Rn\omega}{c} \sin \left(\frac{\omega_0 T}{2} \right) \right]}{\sin \left(\frac{\omega_0 T}{2} \right)} \right\}, \tag{84}$$

$$J_{1b}^{(ij)} = \int_{-\infty}^{\infty} dT \sin \omega T \left\{ \frac{\sin \left[\frac{2Rn\omega}{c} \sin \left(\frac{\omega_0 T}{2} \right) \right]}{\sin \left(\frac{\omega_0 T}{2} \right)} \right\}, \tag{85}$$

$$J_{2A}^{(ij)} = \int_{-\infty}^{\infty} dT \cos \omega_0 T \cos \omega T \left\{ \frac{\sin \left[\frac{2Rn\omega}{c} \sin \left(\frac{\omega_0 T}{2} \right) \right]}{\sin \left(\frac{\omega_0 T}{2} \right)} \right\}, \tag{86}$$

$$J_{2B}^{(ij)} = \int_{-\infty}^{\infty} dT \cos \omega_0 T \sin \omega T \left\{ \frac{\sin \left[\frac{2Rn\omega}{c} \sin \left(\frac{\omega_0 T}{2} \right) \right]}{\sin \left(\frac{\omega_0 T}{2} \right)} \right\}, \tag{87}$$

Using

$$\omega_0 T = \varphi + 2\pi l, \quad \varphi \in (-\pi, \pi), \quad l = 0, \pm 1, \pm 2, ..., \tag{88}$$

we can transform the T-integral into the sum of the telescopic integrals according to the scheme:

$$\int_{-\infty}^{\infty} dT \quad \longrightarrow \quad \frac{1}{\omega_0} \sum_{l=-\infty}^{l=\infty} \int_{-\pi}^{\pi} d\varphi. \tag{89}$$

Using the fact that for the odd functions $f(\varphi)$ and $g(l)$, the relations are valid

$$\int_{-\pi}^{\pi} f(\varphi) d\varphi = 0; \quad \sum_{l=-\infty}^{l=\infty} g(l) = 0, \tag{90}$$

we can write

$$J_{1a}^{(ij)} = \frac{1}{\omega_0} \sum_{l} \int_{-\pi}^{\pi} d\varphi \left\{ \cos \frac{\omega}{\omega_0} \varphi \cos 2\pi l \frac{\omega}{\omega_0} \right\} \left\{ \frac{\sin \left[\frac{2Rn\omega}{c} \sin \left(\frac{\varphi}{2} \right) \right]}{\sin \left(\frac{\varphi}{2} \right)} \right\}, \tag{91}$$

$$J_{1b}^{(ij)} = 0. \tag{92}$$

For integrals with indices A, B we get:

$$J_{2A}^{(ij)} = \frac{1}{\omega_0} \sum_l \int_{-\pi}^{\pi} d\varphi \cos\varphi \left\{ \cos\frac{\omega}{\omega_0}\varphi \cos 2\pi l\frac{\omega}{\omega_0} \right\} \left\{ \frac{\sin\left[\frac{2Rn\omega}{c}\sin\left(\frac{\varphi}{2}\right)\right]}{\sin\left(\frac{\varphi}{2}\right)} \right\}, \qquad (93)$$

$$J_{2B}^{(ij)} = 0, \qquad (94)$$

So, the power spectral formula (80) is of the form:

$$P(\omega,t) = -\frac{\omega}{4\pi^2}\frac{\mu}{n^2}\frac{e^2}{2R} \sum_{i,j=1}^{2} (-1)^{i+j} \left\{ P_1^{(ij)} - n^2\beta^2 P_2^{(ij)} \right\}; \quad \beta = \frac{v}{c}, \qquad (95)$$

where

$$P_1^{(ij)} = J_{1a}^{(ij)} \cos\frac{\omega}{\omega_0}\alpha_{ij} \qquad (96)$$

and

$$P_2^{(ij)} = J_{2A}^{(ij)} \cos\frac{\omega}{\omega_0}\alpha_{ij}. \qquad (97)$$

Using the Poisson theorem

$$\sum_{l=-\infty}^{\infty} \cos 2\pi\frac{\omega}{\omega_0}l = \sum_{k=-\infty}^{\infty} \omega_0\delta(\omega - \omega_0 l), \qquad (98)$$

the definition of the Bessel functions J_{2l} and their corresponding derivations and integrals

$$\frac{1}{2\pi}\int_{-\pi}^{\pi} d\varphi \cos\left(z\sin\frac{\varphi}{2}\right)\cos l\varphi = J_{2l}(z), \qquad (99)$$

$$\frac{1}{2\pi}\int_{-\pi}^{\pi} d\varphi \sin\left(z\sin\frac{\varphi}{2}\right)\sin(\varphi/2)\cos l\varphi = -J_{2l}'(z), \qquad (100)$$

$$\frac{1}{2\pi}\int_{-\pi}^{\pi} d\varphi \frac{\sin\left(z\sin\frac{\varphi}{2}\right)}{\sin(\varphi/2)}\cos l\varphi = \int_0^z J_{2l}(x)dx, \qquad (101)$$

and using equations

$$\sum_{i,j=1}^{2} (-1)^{i+j} \cos\frac{\omega}{\omega_0}\alpha_{ij} = 2(1 - \cos 2\omega t) = 4\sin^2\omega t, \qquad (102)$$

we get with the definition of the partial power spectrum P_l

$$P(\omega) = \sum_{l=1}^{\infty} \delta(\omega - l\omega_0)P_l, \qquad (103)$$

the following final form of the partial power spectrum generated by motion of two-charge system moving in the cyclotron:

$$P_l(\omega, t) = [4(\sin \omega t)^2] \frac{e^2}{\pi n^2} \frac{\omega \mu \omega_0}{v} \left(2n^2 \beta^2 J'_{2l}(2ln\beta) - (1 - n^2\beta^2) \int_0^{2ln\beta} dx\, J_{2l}(x) \right). \quad (104)$$

So we see that the spectrum generated by the system of electron and positron is formed in such a way that the original synchrotron spectrum generated by electron is modulated by function $4\sin^2(\omega t)$. The derived formula involves also the synergic process composed from the synchrotron radiation and the Čerenkov radiation for electron velocity $v > c/n$ in a medium.

Our goal is to apply the last formula in situation where there is a vacuum. In this case we can put $\mu = 1, n = 1$ in the last formula and so we have

$$P_l(\omega, t) = 4\sin^2(\omega t) \frac{e^2}{\pi} \frac{\omega \omega_0}{v} \left(2\beta^2 J'_{2l}(2l\beta) - (1 - \beta^2) \int_0^{2l\beta} dx\, J_{2l}(x) \right). \quad (105)$$

So, we see, that final formula describing the opposite motion of electron and positron in accelerator is of the form

$$P_{l,pair}(\omega, t) = 4\sin^2(\omega t)\, P_{l(electron)}(\omega), \quad (106)$$

where $P_{electron}$ is the spectrum of radiation only of electron. For the same charges it is necessary to replace sine by cosine in the final formula.

The result (106)is surprising because we naively expected that the total radiation of the opposite charges should be

$$P_l(\omega, t) = P_{l(electron)}(\omega, t) + P_{l(positron)}(\omega, t). \quad (107)$$

So, we see that the resulting radiation can not be considered as generated by the isolated particles but by a synergic production of a system of particles and magnetic field. At the same time we cannot interpret the result as a result of interference of two sources because the distance between sources radically changes and so, the condition of an interference is not fulfilled.

The classical electrodynamics formula (106) changes our naive opinion on the electrodynamic processes in the magnetic field. From the last formula it follows that at time $t = \pi k/\omega$ there is no radiation of the frequency ω. The spectrum oscillates with frequency ω. If the radiation were generated not in the synergic way, then the spectral formula would be composed from two parts corresponding to two isolated sources.

5.   The Two Center Circular Motions

The situation which we have analyzed was the ideal situation where the angle of collision of positron and electron was equal to π. Now, the question arises what is the modification of a spectral formula when the collision angle between particles differs from π. It can be easily seen that if the second particle follows the shifted circle trajectory, then the collision angle differs from π. Let us suppose that the center of the circular trajectory of the second particle has coordinates $(a, 0)$. It can be easy to see from the geometry of the situation and from the plane geometry that the collision angle is $\pi - \alpha, \alpha \approx \tan \alpha \approx a/R$ where R is a radius of the first or second circle. The same result follows from the analytical geometry of the situation.

While the equation of the first particle is the equation of the original trajectory, or this is eq. (70)

$$\mathbf{x}_1(t) = \mathbf{x}(t) = R(\mathbf{i} \cos(\omega_0 t) + \mathbf{j} \sin(\omega_0 t)), \tag{108}$$

the equation of a circle with a shifted center is as follows:

$$\mathbf{x}_2(t) = \mathbf{x}(t) = R(\mathbf{i}(\frac{a}{R} + \cos(-\omega_0 t)) + \mathbf{j} \sin(-\omega_0 t)) = \mathbf{x}(-\mathbf{t}) + \mathbf{i}a. \tag{109}$$

The absolute values of velocities of both particles are equal and the relation (72) is valid. Instead of equation (74) we have for radius vectors of particle trajectories:

$$\mathbf{x}_i(t) - \mathbf{x}_j(t') \overset{d}{=} \mathbf{B}_{ij}, \tag{110}$$

where $\mathbf{B}_{11} = \mathbf{A}_{11}, \mathbf{B}_{12} = \mathbf{A}_{12} - \mathbf{i}a, \mathbf{B}_{21} = \mathbf{A}_{21} + \mathbf{i}a, \mathbf{B}_{22} = \mathbf{A}_{22}$.

In general, we can write the last information on coefficients $\mathbf{B}_{ij}$ as follows:

$$\mathbf{B}_{ij} = \mathbf{A}_{ij} + \varepsilon_{ij} \mathbf{i}a, \tag{111}$$

where $\varepsilon_{11} = 0, \varepsilon_{12} = -1, \varepsilon_{21} = 1, \varepsilon_{22} = 0$.

For motion of particles along trajectories the absolute value of vector $\mathbf{A}_{ij} \gg a$ during the most part of the trajectory. It means, we can determine B_{ij} approximatively. After elementary operations, we get:

$$|\mathbf{B_{ij}}| = (A_{ij}^2 + 2|\mathbf{A}_{ij}|\varepsilon_{ij} a \cos \varphi_{ij} + a^2 \varepsilon_{ij}^2)^{1/2}, \tag{112}$$

where $\cos \varphi_{ij}$ can be expressed by the x-component of vector $\mathbf{A}_{ij}$ and $|\mathbf{A}_{ij}|$ as follows:

$$\cos \varphi_{ij} = \frac{(A_{ij})_x}{|\mathbf{A}_{ij}|}. \tag{113}$$

After elementary trigonometric operations, we derive the following formula for $(A_{ij})_x$:

$$(A_{ij})_x = 2R \sin \frac{2\omega_0 t + \omega_0 \tau}{2} \sin \frac{\omega_0 \tau}{2}. \tag{114}$$

Then, using equation (114), we get with $\varepsilon = a/R$

$$|\mathbf{B_{ij}}| = 2R(\sin^2 \frac{\omega_0 \tau + \alpha_{ij}}{2} + \varepsilon \varepsilon_{ij} \sin \frac{2\omega_0 t + \omega_0 \tau}{2} \sin \frac{\omega_0 \tau}{2} + \varepsilon^2 \frac{\varepsilon_{ij}^2}{4})^{1/2}. \tag{115}$$

In order to perform the τ-integration the substitution must be introduced. However, the substitution $\omega_0 \tau + \alpha_{ij} = \omega_0 T$ does not work. So we define the substitution $\tau = \tau(T)$ by the following transcendental equation (we neglect the term with ε^2):

$$\left[\sin^2 \frac{\omega_0 \tau + \alpha_{ij}}{2} + \varepsilon\varepsilon_{ij} \sin \frac{\omega_0 t + \omega_0 \tau}{2} \sin \frac{\omega_0 \tau}{2}\right]^{1/2} = \sin \frac{\omega_0 T}{2}. \tag{116}$$

Or, after some trigonometrical modifications and using the approximative formula $(1+x)^{1/2} \approx 1 + x/2$ for $x \ll 1$

$$[./.]^{1/2} \approx \sin\left(\frac{\omega_0 \tau + \alpha_{ij}}{2}\right) + \frac{\varepsilon}{2}\varepsilon_{ij} \sin\left(\frac{2\omega_0 \tau + 2\omega_0 t - \alpha_{ij}}{2}\right) = \sin \frac{\omega_0 T}{2}. \tag{117}$$

We se that for $\varepsilon = 0$ the substitution is $\omega_0 \tau + \alpha_{ij} = \omega_0 T$. The equation (117) is the transcendental equation and the exact solution is the function $\tau = \tau(T)$. We are looking for the solution of equation (117) in the approximative form using the approximation $\sin x \approx x$. Then, instead of (117) we have:

$$\left(\frac{\omega_0 \tau + \alpha_{ij}}{2}\right) + \frac{\varepsilon}{2}\varepsilon_{ij}\left(\frac{2\omega_0 \tau + 2\omega_0 t - \alpha_{ij}}{2}\right) = \frac{\omega_0 T}{2} \tag{118}$$

Using substitution

$$\omega_0 \tau + \alpha_{ij} = \omega_0 T + \omega_0 \varepsilon A \tag{119}$$

in eq. (118) we get, to the first order in ε-term:

$$A = -\frac{\varepsilon_{ij}}{2\omega_0}(\omega_0 T - 2\alpha_{ij} + 2\omega_0 t). \tag{120}$$

Then, after some algebraic manipulation we get:

$$\omega_0 \tau + \alpha_{ij} = \omega_0 T(1 - \frac{\varepsilon}{2}\varepsilon_{ij}) - \varepsilon\varepsilon_{ij}\omega_0 t(-1)^{i+j} \tag{121}$$

and

$$\omega \tau = \omega T(1 - \frac{\varepsilon}{2}\varepsilon_{ij}) - \frac{\omega}{\omega_0}\left(\varepsilon\varepsilon_{ij}(-1)^{i+j}\omega_0 t + \alpha_{ij}\right). \tag{122}$$

For small time t, we can write approximately:

$$\cos(\omega_0 \tau + \alpha_{ij}) \approx \cos \omega_0 T(1 - \frac{\varepsilon}{2}\varepsilon_{ij}) \tag{123}$$

and from eq. (122)

$$d\tau = dT(1 - \frac{\varepsilon}{2}\varepsilon_{ij}). \tag{124}$$

So, in case of the eccentric circles the formula (118) can be obtained from non-perturbative formula (80) only by transformation

$$T \longrightarrow T(1 - \frac{\varepsilon}{2}\varepsilon_{ij}); \quad \alpha_{ij} \longrightarrow \left(\varepsilon\varepsilon_{ij}(-1)^{i+j}\omega_0 t + \alpha_{ij}\right) = \tilde{\alpha}_{ij}, \tag{125}$$

excepting specific term involving sine functions.

Then, instead of formula (80) we get:

$$P(\omega, t) = -\frac{\omega}{4\pi^2} \frac{e^2}{2R} \frac{\mu}{n^2} \int_{-\infty}^{\infty} dT \sum_{i,j=1}^{2} (-1)^{i+j}$$

$$\times \quad \cos(\omega T - \frac{\omega}{\omega_0}\tilde{\alpha}_{ij}) \left[1 - \frac{c^2}{n^2}v^2 \cos(\omega_0 T)\right] \left\{\frac{\sin\left[\frac{2Rn\omega}{c} \sin\left(\frac{\omega_0 \tilde{T}}{2}\right)\right]}{\sin\left(\frac{\omega_0 \tilde{T}}{2}\right)}\right\}. \tag{126}$$

where $\tilde{T} = T(1 - \frac{\varepsilon}{2}\varepsilon_{ij})$. We see that only $\tilde{T}$ and the α term are the new modification of the original formula (80).

However, because ε term in the sine functions is of very small influence on the behavior of the total function for finite time t, we can neglect it and write approximatively:

$$P(\omega, t) = -\frac{\omega}{4\pi^2} \frac{e^2}{2R} \frac{\mu}{n^2} \int_{-\infty}^{\infty} dT \sum_{i,j=1}^{2} (-1)^{i+j}$$

$$\times \quad \cos(\omega T - \frac{\omega}{\omega_0}\tilde{\alpha}_{ij}) \left[1 - \frac{c^2}{n^2}v^2 \cos(\omega_0 T)\right] \left\{\frac{\sin\left[\frac{2Rn\omega}{c} \sin\left(\frac{\omega_0 T}{2}\right)\right]}{\sin\left(\frac{\omega_0 T}{2}\right)}\right\}. \tag{127}$$

So, we se that only difference with the original radiation formula is in variable $\tilde{\alpha}_{ij}$. It means that instead of sum (102) we have the following sum:

$$\sum_{i,j=1}^{2} (-1)^{i+j} \cos\frac{\omega}{\omega_0}\tilde{\alpha}_{ij} = 2(1 - \cos 2\omega t \cos \varepsilon\omega t). \tag{128}$$

It means that the one electron radiation formula is not modulated by $[\sin \omega t]^2$ but by the formula (128) and the final formula of for the power spectrum is as follows:

$$P_l(\omega, t) = 2(1 - \cos 2\omega t \cos \varepsilon\omega t)P_{l(electron)}(\omega). \tag{129}$$

For $\varepsilon \to 0$, we get the original formula (106).

6. Summary and Discussion

We have derived, in the first part of the chapter, the total quantum loss of energy of the binary. The energy loss is caused by the emission of gravitons during the motion of the two binary bodies around each other under their gravitational interaction. The energy-loss formulas of the production of gravitons are derived here in the source theory. It is evident that the production of gravitons by the binary system is not homogenous and isotropical in space. So, the binary forms the "gravitational light house" where instead of the light photons of the electromagnetic pulsar are the gravitons. The detector of the gravitational waves evidently detects the gravitational pulses.

This section is an extended and revised version of the older author's article (Pardy, 1983a) and preprints (Pardy,1994a,b), in which only the spectral formulas were derived. Here, in the first part of the chapter, we have derived the quantum energy-loss formulas for the linear gravitational field. Linear field corresponds to the weak field limit of the Einstein gravity.

The power spectrum formulas involving radiative corrections are derived in the following part of this chapter, also in the framework of the source theory. The general relativity necessarily does not contain the method how to express the quantum effects together with the radiative corrections by the geometrical language. So, it cannot give the answer on the production of gravitons and on the graviton propagator with radiative corrections. This section therefore deals with the quantum energy loss caused by the production of gravitons and by the radiative corrections in the graviton propagator in case of the motion of a binary.

We believe the situation in the gravity problems with radiative corrections is similar to the QED situation many years ago when the QED radiative corrections were theoretically predicted and then experimentally confirmed for instance in case of he Lamb shift, or, of the anomalous magnetic moment of electron.

Astrophysics is, in a crucial position in proving the influence of radiative corrections on the dynamics in the cosmic space. We hope that the further astrophysical observations will confirm the quantum version of the energy loss of the binary with graviton propagator with radiative corrections.

In the last part of the chapter on pulsars we have derived the power spectrum formula of the synchrotron radiation generated by the electron and positron moving at the opposite angular velocities in homogeneous magnetic field. It forms an analogue of the author article (Pardy, 1997) where only comoving electrons, or positrons was considered, and it forms the modified author preprints (Pardy, 2000a; 2001) and articles (Pardy, 200b; 2002,) where the power spectrum is calculated for two charges performing the retrograde motion in a magnetic field. The frequency of motion was the same because the diameter of the circle was considered the same for both charges. The retrograde motion with different diameters was not considered.

It is surprising that the spectrum depends periodically on radiation frequency ω and time which means that the system composed from electron, positron and magnetic field behaves as a pulsating system. While such pulsar can be represented by a terrestrial experimental arrangement it is possible to consider also the cosmological existence in some modified conditions.

To our knowledge, our result is not involved in the classical monographs on the electromagnetic theory and at the same time it was not still studied by the accelerator experts investigating the synchrotron radiation of bunches. This effect was not described in textbooks on classical electromagnetic field and on the synchrotron radiation. We hope that sooner or later this effect will be verified by the accelerator physicists. The radiative corrections obviously influence the synergistic spectrum of photons (Pardy, 1994c,d).

The particle laboratories used instead of the single electron and positron the bunches with 10^{10} electrons or positrons in one bunch of volume 300μm $\times$ 40μm $\times$ 0.01 m. So, in some approximation we can replace the charge of electron and positron by the charges Q and -Q of both bunches in order to get the realistic intensity of photons. Nevertheless the synergic character of the radiation of two bunches moving at the opposite direction in a magnetic field is conserved.

References

Damour, T.; Taylor, *J. H. Phys. Rev.* **D 45** No. 6 1868 (1992).

Dittrich, W. *Fortschritte der Physik* **26** 289 (1978).

Evans, L. R. *The Large Hadron Collider - Present Status and Prospects* CERN-OPEN-99-332 CERN Geneva (1999).

Gold, T. *Nature* **218** 731 (1968).

Goldreich, P.; Julian, W. H. *Astrophys. J.* **157** 869 (1969).

Graham-Smith, F. *Rep. Prog. Phys.* **66** 173 (2003).

Hewish, A.; Bell, S. J.; Pilkington, J. D. H. ; et al. *Nature* **217** 709 (1968).

Huguenin, G. R; Taylor, J. H.; Goad, L. E.; Hartai, A.; Orsten, G. S. F.; Rodman, A. K. *Nature* **219** 576 (1968).

Hulse, R. A.; Taylor, J. H. Astrophys. *J. Lett.,* **195** L51-L53 (1975).

Cho C. F. and Harri Dass, N. D. *Ann. Phys. (NY)* **90** 406 (1976).

Landau, L. D. Phys. *Zs. Sowjet* **1** 285 (1932).

Manchester, R. N. *Phil. Trans. R. Soc. Lond.* **A 341** 3 (1992).

Manoukian, E. B. *GRG* **22** 501 (1990).

Melrose, D. B. *Phil. Trans. R. Soc. Lond.* **A 341** 105 (1992).

Pardy, M. *GRG* **15** No. 11 1027 (1983a).

Pardy, M. *Phys. Lett.* **94A** 30 No. 1 (1983b).

Pardy, M. *Phys. Lett.* **140A** 51 Nos. 1,2 (1989).

Pardy, M. CERN-TH.7239/94 (1994a).

Pardy, M. CERN-TH.7299/94 (1994b).

Pardy, M. *Phys. Lett.* **B 325** 517 (1994c).

Pardy, M. *Phys. Lett.* **A 189** 227 (1994d).

Pardy, M. *Phys. Lett.* **B 336** 362 (1994e).

Pardy, M. *Phys. Rev.* **A 55** No. 3 1647 (1997).

Pardy, M. hep-ph/0001277 (2000a).

Pardy, M. *Int. Journal of Theor. Phys.* **39** No. 4 1109 (2000b).

Pardy, M. hep-ph/011036 (2001).

Pardy, M. *Int. Journal of Theor. Phys.* **41** No. 6 1155 (2002).

Seiradakis, J. H.; Wielebinski, R. *Astronomy & Astrophysics Review* manuscript (2004), hep-ph/0410022 (2004).

Schott, G. A. *Electromagnetic Radiation* (Cambridge University Press, 1912).

Schwinger, J. *Phys. Rev.* **75** 1912 (1949).

Schwinger, J. *GRG* **7** No. 3 251 (1976).

Schwinger, J. *Particles, Sources and Fields*, Vol. I, (Addison-Wesley, Reading, Mass., 1970).

Schwinger, J. *Particles, Sources and Fields*, Vol. II, (Addison-Wesley, Reading, Mass., 1973).

Schwinger, J. *Phys. Rev.* **75** 1912 (1949).

Schwinger, J.; Tsai, W. Y; Erber, T. *Ann. Phys. (NY)* **96** 303 (1976).

Sokolov, A. A.; Ternov, I. M. *The relativistic electron* (Moscow, Nauka, 1983). (in Russian).

Taylor, J. H.; Wolszczan, A.; Damour, T.; Weisberg, J. M. *Nature* **355** 132 (1992).

Taylor, J. H. Jr. *Binary Pulsars and Relativistic Gravity* (Nobel Lecture, 1993).

Weinberg, S. *Gravitation and Cosmology* (John Wiley and Sons, Inc., New York, 1972).

In: Pulsars: Discoveries, Functions and Formation
Editor: Peter A. Travelle, pp. 125-160

ISBN: 978-1-61122-982-0
© 2011 Nova Science Publishers, Inc.

Chapter 7

PARTICLE ACCELERATION IN PULSAR OUTER MAGNETOSPHERES: ELECTRODYNAMICS AND HIGH-ENERGY EMISSIONS

Kouichi Hirotani[*]
ASIAA/National Tsing Hua University - TIARA,
PO Box 23-141, Taipei, Taiwan

Abstract

In the last 15 years our knowledge on the high-energy emission from rotation-powered pulsars has considerably improved. The seven known gamma-ray pulsars provide us with essential information on the properties of the particle accelerator — electro-static potential drop along the magnetic field lines. Although the accelerator theory has been studied for over 30 years, the origin of the pulsed gamma-rays is an unsettled question. In this chapter, we review accelerator theories, focusing on the electrodynamics of outer-magnetospheric gap models. Then we present a modern accelerator theory, in which the Poisson equation for the electro-static potential is solved self-consistently with the Boltzmann equations for electrons, positrons, and gamma-rays.

Keywords: gamma-rays: observations, gamma-rays: theory, magnetic fields, pulsars: individual, (Crab pulsar)

1. Introduction

The Energetic Gamma Ray Experiment Telescope (EGRET) on board the *Compton Gamma-ray Observatory* (CGRO) has detected pulsed signals from at least six rotation-powered pulsars (for the Crab pulsar, Nolan et al., 1993, Fierro et al., 1998; for the Vela pulsar, Kanbach et al., 1994, Fierro et al., 1998; for PSR B1706–44, Thompson et al., 1996; for PSR B1951, Ramanamurthy et al., 1995; for Geminga, Mayer-Hasselwander et

---

[*]E-mail address: hirotani@tiara.sinica.edu.tw; Postal address: TIARA, Department of Physics, National Tsing Hua University, 101, Sec. 2, Kuang Fu Rd., Hsinchu, Taiwan 300

al., 1994, Fierro et al., 1998; for PSR B1055–52, Thompson et al., 1999). All of them show a double peak in their light curves. At lower energies (below 10 MeV), COMPTEL aboard CGRO also detected pulsation from PSR B1509–58 (Kuiper et al., 1999). In addition to these seven high-confidence pulsar detections above MeV, at least three radio pulsars may have been seen by EGRET (PSR B1046–58, Kaspi at al., 2000; PSR B0656+14, Ramanamurthy et al., 1996; PSR J0218+4232, Kuiper et al., 2000). Since interpreting γ-rays should be less ambiguous compared with reprocessed, non-thermal X-rays, the γ-ray pulsations observed from these objects are particularly important as a direct signature of basic non-thermal processes in pulsar magnetospheres, and potentially should help to discriminate among different emission models.

Spectral energy distributions (SEDs) offers the key to an understanding of the radiation processes. Thompson (1999) compiled a useful set of broad-band SED for the seven high-confidence γ-ray pulsars. The most striking feature of these νF_ν plots is the flux peak above 0.1 GeV with a turnover at several GeV. Various models (Daugherty & Harding 1982, 1996a,b, hereafter DH82, DH96a,b; Romani and Yadigaroglu 1995, hereafter RY95; Romani 1996, hereafter R96; Zhang & Cheng 1997, hereafter ZC97; Higgins & Henriksen 1997, 1998; Hirotani & Shibata 1999a, b, c, hereafter Papers I, II, III) conclude that these photons are emitted by the electrons or positrons accelerated above 5 TeV via curvature process. Such ultra-relativistic particles could also cause inverse-Compton (IC) scatterings. In particular, for young pulsars, their strong thermal X-rays emitted from the cooling neutron-star surface (Becker& Trümper 1997; Pavlov et al., 2001; Kaminker et al., 2002) efficiently illuminate the gap to be upscattered into several TeV energies. If their magnetospheric infrared (0.1–0.01 eV) photon field is not too strong, a pulsed IC component could be unabsorbed to be detected with future ground-based or space telescopes.

The curvature-emitted photons have the typical energy of a few GeV. Close to the star (typically within a few stellar radii), such γ-rays can be absorbed by the strong magnetic field ($> 10^{12}$ G) to materialize as a pair. On the other hand, outside of this strong-field region, pairs are created only by the photon-photon collisions (e.g., between the curvature-GeV photons and surface- or magnetospheric- keV photons). The replenished charges partially screen the original acceleration field, $E_\parallel$. If the created particles pile up at the boundaries of the potential gap, they will quench the gap eventually. Nevertheless, if the created particles continue to migrate outside of the gap as a part of the global flows of charged particles, a steady charge-deficient region will be maintained. This is the basic idea of a particle acceleration zone in a pulsar magnetosphere.

The pulsar magnetosphere can be divided into two zones: The closed zone filled with a dense plasma corotating with the star, and the open zone in which plasma flows along the open field lines to escape through the light cylinder (left panel of fig. 1). These two zones are separated by the the last-open magnetic field lines, which become parallel to the rotation axis at the light cylinder. Here, the light cylinder is defined as the surface where the corotational speed of a plasma would coincide with the speed of light, c, and hence its distance from the rotation axis is given by $\varpi_{\mathrm{LC}} \equiv c/\Omega$. If a plasma flows along the magnetic field line, causality requires that the plasma should migrate outward outside of the light cylinder. In all the pulsar emission models, particle acceleration takes place within the open zone.

For an aligned rotator (i.e., magnetic inclination angle, α_i, vanishes), open zone occu-

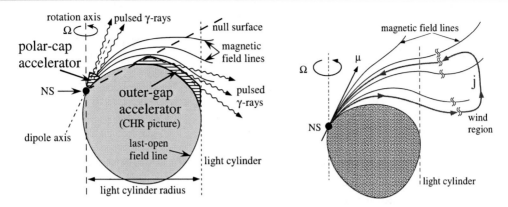

Figure 1. Sideview of pulsar magnetosphere. *Left:* Two representative accelerator models in the open zone. The small filled circle represents the neutron star. On the null surface (heavy dashed line),the magnetic field component projected along the rotation axis, vanishes. *Right:* Global current induced by the EMF exerted on the spinning neutron star surface.

pies the magnetic colatitudes (measured from the magnetic axis) that is less than $\sqrt{r_*/\varpi_{LC}}$, where r_* represents the neutron-star radius. For an oblique rotator, even though the open-zone polar cap shape is distorted, $\pi(r_*/\varpi_{LC})$ gives a good estimate of the polar cap area. On the spinning neutron star surface, from the magnetic pole to the rim of the polar cap, an electro-motive force (EMF) $\approx \Omega^2 B_* r_*^3/c^2 \approx 10^{16.5}$ V, is exerted, where B_* refers to the polar-cap magnetic field strength. In this chapter, we assume that both the spin and magnetic axes reside in the same hemisphere; that is, $\vec{\Omega} \cdot \vec{\mu} > 0$, where $\vec{\Omega}$ represents the rotation vector, and $\vec{\mu}$ the stellar magnetic moment vector. In this case, the strong EMF induces the magnetospheric currents that flow outwards in the lower latitudes and inwards along the magnetic axis (right panel in fig. 1). The return current is formed at large-distances where Poynting flux is converted into kinetic energy of particles or dissipated (Shibata 1997).

Attempts to model the pulsed γ-ray emissions have concentrated on two scenarios (fig. 1): Polar cap models with emission altitudes of $\sim 10^4$cm to several neutron star radii over a pulsar polar cap surface (Harding, Tademaru, & Esposito 1978; DH82; DH96a,b; Dermer & Sturner 1994; Sturner, Dermer, & Michel 1995; for the slot gap model also see Scharlemann, Arons, & Fawley 1978, hereafter SAF78), and outer gap models with acceleration occurring in the open field zone located near the light cylinder (Cheng, Ho, & Ruderman 1986a,b, hereafter CHR86a,b; Chiang & Romani 1992, 1994; R96; RY95). Both models predict that electrons and positrons are accelerated in a charge depletion region, a potential gap, by the electric field along the magnetic field lines to radiate high-energy γ-rays via the curvature process. However, it is worth noting that average location of energy loss should take place near the light cylinder so that the rotating neutron star may lose enough angular momentum (Shibata 1995), which indicates the existence of the polar-cap accelerator must affect the electrodynamics in the outer magnetosphere.

It is widely accepted from phenomenological studies that coherent *radio* pulsations are probably emitted from the polar-cap accelerator. Moreover, coherent radio pulsations provide the primary channel for pulsar discovery (1627 pulsars as of June 2006, see ATNF

pulsar catalogue, http://www.atnf.csiro.au/research/pulsar/psrcat/). However, it is, unfortunately, very difficult to reproduce them by considering plasma collective effects. Therefore,
in this chapter, I will focus on *incoherent high-energy*, non-thermal radiation, which is
likely the product of the potential gap in the outer magnetosphere.

2. Traditional Outer-gap Model

For a static observer, electric field $\vec{E}$ is obtained from the Poisson equation, $\vec{\nabla} \cdot \vec{E} = 4\pi\rho$,
where ρ denotes the real charge density. Since only the electric field component projected
along the magnetic field line contributes for particle acceleration, we decouple $\vec{E}$ into the
parallel $\vec{E}_{\parallel}$ and perpendicular $\vec{E}_{\perp}$ components with respect to the magnetic field line to
obtain

$$\vec{\nabla} \cdot \vec{E}_{\parallel} = 4\pi(\rho - \rho_{GJ}), \tag{1}$$

where

$$\vec{E}_{\parallel} = -\vec{\nabla}\Psi \equiv -\vec{\nabla}(A_t + \Omega A_\varphi) \tag{2}$$

and (Goldreich & Julian 1969; Mestel 1971)

$$\rho_{GJ} \equiv \frac{\vec{\nabla} \cdot \vec{E}_{\perp}}{4\pi} = -\frac{\vec{\Omega} \cdot \vec{B}}{2\pi c} + \frac{(\vec{\Omega} \times \vec{r}) \cdot (\vec{\nabla} \times \vec{B})}{4\pi c}; \tag{3}$$

A_t and A_φ represent the scalar potential and magnetic flux function, respectively, $\vec{B}$ the
magnetic field vector at position $\vec{r}$ from the stellar center. For more rigorous derivation
of the Poisson equation for arbitrary magnetic field configuration in the curved spacetime,
see § 4.1.. For a Newtonian dipole magnetic field, ρ_{GJ} changes sign at the null surface
(heavy dashed line in fig. 1) on which B_ζ, the magnetic field component projected along
the rotation axis, vanishes. If ρ deviates from ρ_{GJ} in any region, an electric field is exerted
along $\vec{B}$. If the potential drop is sufficient, migratory electrons and/or positrons will be
accelerated to radiate γ-rays via curvature and/or inverse-Compton (IC) processes.

To examine the gap electrodynamics, it is convenient to adopt the so-called 'magnetic
coordinates' (s, θ_*, φ_*) such that s denotes the distance along the field line, θ_* and φ_* the
magnetic colatitude and the magnetic azimuth of the point where the field line intersects
the stellar surface; $\theta_* = 0$ and $\varphi_* = 0$ correspond to the magnetic axis and to the plane on
which both the rotation and the magnetic axes reside, respectively. If we assume that the
gap is thin in the meridional direction in the sense that the θ_* derivatives dominate the s and
φ_* derivatives in the left-hand side of equation (1), we obtain (SAF78; see also Muslimov
& Tsygan 1992 for the general-relativistic expression)

$$-\frac{1}{r^2}\left(\frac{\partial\theta_*}{\partial\theta}\right)^2_{\varphi,r} \frac{1}{\theta_*}\frac{\partial}{\partial\theta_*}\left(\theta_*\frac{\partial\Psi}{\partial\theta_*}\right) = -\frac{1}{r_*^2}\frac{B}{B_*}\frac{1}{\theta_*}\frac{\partial}{\partial\theta_*}\left(\theta_*\frac{\partial\Psi}{\partial\theta_*}\right) = 4\pi(\rho - \rho_{GJ}). \tag{4}$$

Neglecting the θ_* dependence of the effective charge $\rho_{\text{eff}} \equiv \rho - \rho_{GJ}$, we obtain

$$\Psi \approx \frac{\Omega B_*}{c}\left(\frac{\rho}{\Omega B/2\pi c} - \frac{\rho_{GJ}}{\Omega B/2\pi c}\right)r_*^2(\theta_* - \theta_*^{\min})(\theta_*^{\max} - \theta_*), \tag{5}$$

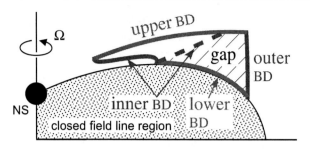

Figure 2. Definition of boundaries of outer gap. The inner boundary is located at the null surface (dashed line) when the gap is vacuum as in CHR picture, but shifts inwards as indicated by the solid curve as the gap becomes non-vacuum (see text).

where θ_*^{max} and θ_*^{min} give the lower and upper boundary positions, respectively (fig. 2). We should notice here that in traditional outer-gap model, it is assumed $|\rho| \ll \rho_{\mathrm{GJ}}$. Thus, $\rho_{\mathrm{GJ}}/B \propto B_\zeta/B$ essentially determines the s dependence of Ψ and hence $E_\parallel \equiv -\partial_s \Psi$. Since ρ_{GJ}/B is roughly proportional to $s - s_0$ (at least near the null surface, $s = s_0$), one obtains a roughly constant $E_\parallel$ in the traditional outer-gap model. The strength of $E_\parallel$ is determined by the Goldreich-Julian charge density at the surface, $\Omega B_*/2\pi c$. Moreover, $E_\parallel(s, \theta_*)$ has a quadratic dependence on θ_* because of the two zero-potential walls at $\theta_* = \theta_*^{\mathrm{max}}$ and θ_*^{min}. By solving equation (1) on the two-dimensional plane (s, θ_*), we can confirm later that $\Psi > 0$ holds for $s < s_0$ and $\Psi < 0$ for $s > s_0$ provided that the gap is vacuum (i.e., $\rho = 0$) and geometrically thin (i.e., $\theta_*^{\mathrm{max}} - \theta_*^{\mathrm{min}} \ll 1$). Therefore, a vacuum, thin gap starts from the null surface and extends towards the light cylinder. The region $0 < s < s_0$ will be filled with the electrons extracted from the stellar surface, $s = 0$, and hence in-active. Strictly speaking, because of the non-zero contribution of ρ (and also of the two-dimensional effect), the inner boundary $s = s_{\mathrm{in}}$ of such a gap is located slightly inside of $s = s_0$.

In traditional outer-gap models, pulsar high-energy properties have been argued, assuming that the (dimensionless) created current density per magnetic flux tube, $j_{\mathrm{gap}} = -2\pi c \rho(s_{\mathrm{in}})/\Omega B(s_{\mathrm{in}})$ becomes comparable to the typical Goldreich-Julian value, $B_{\zeta*}/B_*$, where $B_{\zeta*}$ represents B_ζ at the stellar surface. Note that $j_{\mathrm{gap}} \sim B_{\zeta*}/B_*$ is necessary to obtain the observed γ-ray fluxes and that only electrons (with velocity $-c$) exist at $s = s_{\mathrm{in}}$. However, the non-vanishing $\rho(s_{\mathrm{in}}) = -j_{\mathrm{gap}}\Omega B/2\pi c$ cannot be neglected in the right-hand side of equation (1) at $s = s_0$, where ρ_{GJ} vanishes. This fact shows that the traditional thought that $s_{\mathrm{in}} \approx s_0$ breaks down for non-vacuum gaps.

Let us consider the inner boundary position under non-vanishing created charges. In the outer part of the gap, ρ/B tends to be constant because of the decreasing pair creation. Thus, $-\partial_s(\rho_{\mathrm{eff}}/B) \approx \partial_s(\rho_{\mathrm{GJ}}/B) > 0$ leads to a positive $E_\parallel$ in a transversely thin gap (eq. [5]). In order not to change the sign of $E_\parallel$, $\partial_s E_\parallel \approx 4\pi(\rho - \rho_{\mathrm{GJ}})$ should be positive in the vicinity of the inner boundary (eq. [1]). Thus, we obtain

$$-\frac{\rho_{\mathrm{GJ}}}{B} \sim \frac{\Omega}{2\pi c}\frac{B_\zeta}{B} > \frac{-\rho}{B} = j_{\mathrm{gap}}\frac{\Omega}{2\pi c}, \tag{6}$$

at $s = s_{\mathrm{in}}$. It follows that the polar cap is the only place for the inner boundary of the 'outer' gap to be located, if j_{gap} is as large as $B_{\zeta*}/B_*$. (On the other hand, if $j_{\mathrm{gap}} \ll B_{\zeta*}/B_*$ holds, the inner boundary is located near the null surface, where B_ζ vanishes.) Such a non-vacuum gap must extend from the polar cap surface (not from the null surface as traditionally assumed) to the outer magnetosphere. We can therefore conclude that the original vacuum solution obtained by CHR86a,b cannot be applied to a non-vacuum gap, which is necessary to explain the observed γ-ray fluxes. To construct a self-consistent model, we have to solve equation (1) together with the Boltzmann equations for particles and γ-rays. This scheme was first proposed by Beskin et al., (1992, hereafter BIP92) for a magnetosphere of a super-massive black hole (SMBH), and developed by Hirotani and Okamoto (1997, hereafter Paper 0), then applied to pulsar magnetospheres (Papers I, II, III). In what follows, we review this method of approach.

For convenience, we summarize accelerator models in table 1. The three-dimensional model by RY95, R96, and Cheng et al. (2000, hereafter CRZ00) bases on CHR86a,b. For comparison, we also present representative inner-gap models (Fawley et al., 1977, hereafter FAS77; SAF78; DH96a, b) in the table.

3.   1-Dimensional Analysis of Gap Electrodynamics

A good place to start is the one-dimensional consideration of gap properties along the magnetic field lines. Neglecting relativistic effects, and assuming that typical transfield thickness, $W_\perp$, is greater than or comparable to the longitudinal width, $W_\|$, we can reduce the Poisson equation for non-corotational potential Ψ into the one-dimensional form

$$-\nabla^2\Psi \approx -\frac{d^2\Psi}{ds^2} + \frac{\Psi}{W_\perp^2} = 4\pi\left[\rho(s) + \frac{\Omega B_\zeta(s)}{2\pi c}\right]. \tag{7}$$

The real charge density can be computed by

$$\rho = e\int_1^\infty (N_+ - N_-)d\Gamma, \tag{8}$$

where $N_+(s,\Gamma)$ and $N_-(s,\Gamma)$ refer to the positronic and electronic distribution functions, respectively.

3.1.   Particle Boltzmann Equations

In general, the distribution functions $N_\pm = N_\pm(t, \vec{x}, \vec{p})$ obey the following Boltzmann equations:

$$\frac{\partial N_\pm}{\partial t} + \frac{\vec{p}}{m_e\Gamma}\cdot\vec{\nabla}N_\pm + \vec{F}_{\mathrm{ext}}\cdot\frac{\partial N_\pm}{\partial\vec{p}} = S_\pm(t, \vec{x}, \vec{p}), \tag{9}$$

where $\vec{p}$ refers to the particle momentum, $\vec{F}_{\mathrm{ext}}$ the external forces acting on particles, and $S_\pm$ the collision terms. In this chapter, $\vec{F}_{\mathrm{ext}}$ consists of the Lorentz and the curvature radiation reaction forces. Since the magnetic field is much less than the critical value (4.41×10^{13} G), quantum effects can be neglected in the outer magnetosphere. As a result, curvature radiation takes place continuously and can be regarded as an external force acting on a particle. If we instead put the collision term associated with the curvature process in the right-hand side, the energy transfer in each collision would be too small to be resolved by the energy grids. We take the γ-ray emission rate due to curvature process into account consistently in the γ-ray Boltzmann equations. Throughout this chapter, we assume a stationary magnetosphere (on the corotating frame) and impose

$$\frac{\partial}{\partial t} + \Omega \frac{\partial}{\partial \phi} = 0. \tag{10}$$

Collision terms are expressed as

$$S_\pm(s,\theta_*,\varphi_*,\Gamma,\chi) = -\int_{-1}^{1} d\mu_c \int_{E_\gamma < \Gamma} dE_\gamma \eta_{IC}^\gamma(E_\gamma,\Gamma,\mu_c) n_\pm(s,\theta_*,\varphi_*,\Gamma,\chi)$$

$$+ \int_{-1}^{1} d\mu_c \int_{\Gamma_i > \Gamma} d\Gamma_i \, \eta_{IC}^e(\Gamma_i,\Gamma,\mu_c) n_\pm(s,\theta_*,\varphi_*,\Gamma_i,\chi)$$

$$+ \int_{-1}^{1} d\mu_c \int dE_\gamma \left[\left(\frac{\partial \eta_{\gamma\gamma}(E_\gamma,\Gamma,\mu_c)}{\partial \Gamma} + \frac{\partial \eta_{\gamma B}(E_\gamma,\Gamma,\mu_c)}{\partial \Gamma} \right) \frac{B_*}{B} g_\pm(r,E_\gamma,\vec{k}) \right], \tag{11}$$

where μ_c refers to the cosine of the collision angle between the particle and the soft photon for inverse-Compton scatterings (ICS), between the γ-ray and the soft photon for two-photon pair creation, and between the γ-ray and the local magnetic field line for one-photon pair creation; χ denotes the pitch angle of the particle. The function g represents the γ-ray distribution function divided by $\Omega B_*/(2\pi ce)$ at energy E_γ, momentum $\vec{k}$ and position $\vec{r}$. In the third line, μ_c is computed from the photon propagation direction, $\vec{k}/|\vec{k}|$. Since pair annihilation is negligible, we do not include this effect in equation (11). For the explicit expressions of the re-distribution functions, η_{IC}^γ, η_{IC}^e, $\eta_{\gamma\gamma}$, and $\eta_{\gamma B}$, see § 4.2. (see also Hirotani 2006a, hereafter Paper XI).

For the sake of one-dimensional analysis, in this section, we suppress the dependence on θ_*, φ_*, and χ in equations (9) and (11) to elucidate basic features of gap electrodynamics in the one-dimensional approximation. To suppress the dependence of g on $\vec{k}/|\vec{k}|$, we also assume that γ-rays propagate parallel to the local magnetic field line and adopt $\vec{k} \parallel \vec{B}$. Imposing stationary condition (10), we then obtain

$$\pm c \frac{\partial n_\pm}{\partial s} + \frac{1}{m_e c} \left(eE_\parallel - \frac{P_{SC}}{c} \right) \frac{\partial n_\pm}{\partial \Gamma}, = S_\pm. \tag{12}$$

where

$$S_\pm(s,\Gamma) = -\int_{E_\gamma < \Gamma} dE_\gamma \eta_{IC}^\gamma(E_\gamma,\Gamma,\mu_c) n_\pm(s,\Gamma)$$

$$+ \int_{\Gamma_i > \Gamma} d\Gamma_i \, \eta_{IC}^e(\Gamma_i,\Gamma,\mu_c) n_\pm(s,\Gamma_i) + \int dE_\gamma \frac{\partial \eta_{\gamma\gamma}(E_\gamma,\Gamma,\mu_c)}{\partial \Gamma} \frac{B_*}{B} g_\pm(\vec{r},E_\gamma). \tag{13}$$

Note that μ_c is determined at each position s. In the first and second terms in the right-hand side, we assume the surface blackbody component as the source of the photons to be up-scattered and adopt uni-directional approximation to their specific intensity. In the third line, $\vec{k} \parallel \vec{B}$ determines μ_c at each point.

3.2. Gamma-ray Boltzmann Equations

In general, the γ-ray distribution function g obeys the Boltzmann equation,

$$\frac{\partial g}{\partial t} + c \frac{\vec{k}}{|\vec{k}|} \cdot \nabla g(t, \vec{r}, \vec{k}) = S_\gamma(t, \vec{r}, \vec{k}), \tag{14}$$

where $|\boldsymbol{k}|^2 \equiv -k^i k_i$; S_γ is given by

$$
\begin{aligned}
S_\gamma = & -\int_{-1}^{1} d\mu_c \int_{1}^{\infty} d\Gamma \frac{\partial [\eta_{\gamma\gamma}(\vec{r}, \Gamma, \mu_c) + \eta_{\gamma B}(\vec{r}, \Gamma, \mu_c)]}{\partial \Gamma} \cdot g(\vec{r}, E_\gamma, \vec{k}) \\
& + \int_{-1}^{1} d\mu_c \int_{1}^{\infty} d\Gamma \eta_{IC}^\gamma(E_\gamma, \Gamma, \mu_c) \frac{B}{B_*} n_\pm(s, \theta_*, \varphi_*, \Gamma, \chi) \\
& + \int_{0}^{\pi} d\chi \int_{1}^{\infty} d\Gamma \eta_{SC}(E_\gamma, \Gamma, \chi) \frac{B}{B_*} n_\pm(s, \theta_*, \varphi_*, \Gamma, \chi), \tag{15}
\end{aligned}
$$

where η_{SC} is the synchro-curvature radiation rate [s^{-1}] into the energy interval between E_γ and $E_\gamma + dE_\gamma$ by a particle migrating with Lorentz factor Γ and pitch angle χ. For explicit expression, see Cheng and Zhang (1996).

Imposing the stationary condition (10), and denoting $g_\pm$ as the g associated with the γ-rays propagating in $\pm \vec{B}$ directions, we obtain the following Boltzmann equations:

$$\pm c \frac{\partial g_\pm}{\partial s} = S_{\gamma, \pm}(s, E_\gamma), \tag{16}$$

where

$$
\begin{aligned}
S_{\gamma, \pm} = & -\int_{1}^{\infty} d\Gamma \frac{\partial \eta_{\gamma\gamma}(s, \Gamma, \mu_c)}{\partial \Gamma} \cdot g_\pm(s, E_\gamma) \\
& + \int_{1}^{\infty} d\Gamma \eta_{IC}^\gamma(E_\gamma, \Gamma, \mu_c) \frac{B}{B_*} n_\pm(s, \Gamma) + \int_{1}^{\infty} d\Gamma \eta_{SC}(E_\gamma, \Gamma, \chi) \frac{B}{B_*} n_\pm(s, \Gamma), \tag{17}
\end{aligned}
$$

3.3. Boundary Conditions

The Poisson equation and the γ-ray Boltzmann equations are ordinary differential equations, which can be straightforwardly solved by a simple discretization. On the other hand, the hyperbolic-type partial differential equations (12) are solved by the Cubic Interpolated Propagation (CIP) method (e.g., Yabe & Aoki 1991, Yabe, Xiao, & Utsumi 2001). We represent $n_\pm$ at $\Gamma = \beta_l$ ($l = 1, 2, 3, \ldots, m$) with $n_{\pm, l}$, where β_l is a linearly gridded Lorentz factor variable.

At the *inner* (starward) boundary ($s = s_{in}$), we impose (Hirotani & Shibata 2001a, b, hereafter Papers VII, VIII)

$$E_\parallel(s_{in}) = 0, \quad \Psi(s_{in}) = 0, \tag{18}$$

$$g_+^i(s_{in}) = 0 \quad (i = l_0 + 1, \ldots, -1, 0, 1, 2, \ldots, m), \tag{19}$$

$$n_{+, l}(s_{in}) = \begin{cases} j_{in}, & \text{for } l = 1 \\ 0, & \text{for } l = 2, 3, \ldots, m \end{cases} \tag{20}$$

where the dimensionless positronic injection current across the inner boundary $s = s_{in}$ is denoted as j_{in}. In this section, we adopt $\beta_0 = 10^5$. Current conservation gives another constraint

$$\sum_l n_{-,l}(s_{in}) = j_{tot} - j_{in}. \qquad (21)$$

At the *outer* boundary ($s = s_{out}$), we impose

$$E_\parallel(s_{out}) = 0, \qquad (22)$$

$$g_-^i(s_{out}) = 0 \qquad (i = l_0 + 1, \ldots, -1, 0, 1, 2, \ldots, m), \qquad (23)$$

$$n_{-,l}(s_{out}) = \begin{cases} j_{out}, & \text{for } l = 1 \\ 0, & \text{for } l = 2, 3, \ldots, m \end{cases} \qquad (24)$$

The current density created in the gap per unit flux tube can be expressed as

$$j_{gap} = j_{tot} - j_{in} - j_{out}. \qquad (25)$$

We adopt j_{gap}, j_{in}, and j_{out} as the free parameters.

We have totally $2m + 2(m - l_0 + 1) + 4$ boundary conditions (18)–(24) for $2m + 2(m - l_0 + 1) + 2$ unknown functions $n_{\pm l}$, $g_{\pm i}$, Ψ, and $E_\parallel$. Thus two extra boundary conditions must be compensated by making the positions of the boundaries s_{in} and s_{out} be free. The two free boundaries appear because $E_\parallel = 0$ is imposed at *both* the boundaries and because j_{gap} is externally imposed. In other words, the gap boundaries (s_{in} and s_{out}) shift, if j_{in} and/or j_{out} varies.

3.4. Mono-energetic Approximation: Acceleration Electric Field

To aid in grasping the basic features, we first adopt the mono-energetic approximation in which Γ dependence of $n_\pm(s, \Gamma)$ is suppressed, assuming that all the particles have the Lorentz factor that is given by the balance between the electro-static acceleration and the curvature radiation-reaction force. Under this simplification, all the basic equations (7), (12), and (16) become ordinary differential equations. Note that $m = 1$ is adopted (see § 3.3.).

Let us begin with browsing some examples of $E_\parallel(s)$ solved for several representative values of the created current density per magnetic flux tube, j_{gap}. In figure 3, we present $E_\parallel(s)$ and the equilibrium Lorentz factor, which is obtained by equating P_{SC}/c with $eE_\parallel$, for $j_{gap} = 0.10, 0.20, 0.218$ with solid, dashed, and dotted curves, respectively (Paper III). Other parameters are chosen to be $\Omega = 100 \text{rad s}^{-1}$, $\mu = 10^{30}\text{G cm}^3$, $W_\perp = 10^8$ cm, $kT = 75$ eV (surface blackbody temperature), and $\alpha_i = 0°$ (i.e., aligned rotator). In this figure, abscissa denotes the distance along the field line, $x = s - s_0$.

For a small j_{gap}, the gap becomes nearly vacuum. Moreover, unless the gap width, $W_\parallel \equiv s_{out} - s_{in}$, exceeds $W_\perp$, the term $\Psi/W_\perp^2$, which describes the 2-dimensional screening effect, is not important. As a result, equation (7) gives approximately a quadratic solution,

$$E_\parallel(s) = E_\parallel(s_0) - \frac{\Omega}{c}\left(\frac{\partial B_\zeta}{\partial s}\right)_0 (s - s_0)^2. \qquad (26)$$

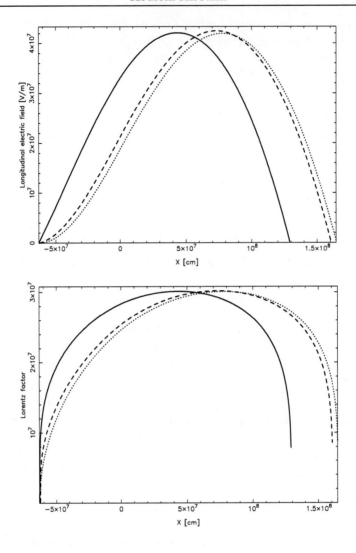

Figure 3. Solution for $j_{\mathrm{gap}} = 0.10$ (solid), 0.20 (dashed), 0.218 (dotted). *Top*: The field-aligned electric field; *Bottom*: equilibrium Lorentz factor. Both panels are from Paper III.

   The solid line in the left panel of figure 3 (i.e., small j_{gap} case) approximately represents such a quadratic solution. Since the trans-field thickness ($= 10^8$cm) is comparable with the longitudinal gap width in the present case, the two-dimensional effect contributes to shift the gap outwards. In another word, without the $\Psi/W_\perp^2$ term, the solid curve would become symmetric with respect to $x = s - s_0 = 0$.

   It is worth examining the case when $W_\perp$ becomes smaller. In figure 4, we present $E_\parallel(s)$ for three discrete values of $W_\perp$ (Paper I). Other quantities are fixed as $j_{\mathrm{gap}} = 0.1$, $\Omega = 100\mathrm{rad\,s}^{-1}$ and $\alpha_{\mathrm{i}} = 0°$. Instead of specifying μ and kT, we adopt $B = 10^5$ G at $s = s_0$ and $L = 10^{33}\mathrm{ergs\,s}^{-1}$ for the surface blackbody luminosity. It follows that $E_\parallel(s)$ symmetrically distribute with respect to the null surface when $W_\perp$ is small and that the gap shift outwards with decreasing $W_\perp$ because of the two-dimensional screening effect. It suggests that a vacuum, transversely thin (i.e., $j_{\mathrm{gap}} \ll 1$ and $W_\perp \ll W_\parallel$) gap starts from the null surface to extend towards the light cylinder with a nearly constant $E_\parallel$. It will be confirmed by a full two-dimensional analysis in § 4.5.1..

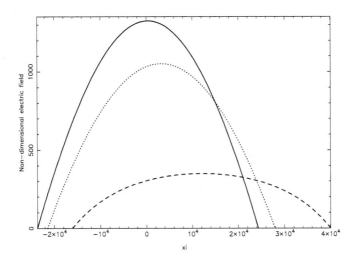

Figure 4. Solution of $E_\parallel(\xi)$ for $W_\perp = 1.0 \times 10^8$ cm (solid), 0.3×10^8 cm (dashed), and 0.1×10^8 cm (dotted). The actual distance along the field line is given by $s - s_0 = 1.2 \times 10^3 \xi$ cm. From Paper I.

Let us examine how $E_\parallel(s)$ changes with particle injection. In figure 5, we present the case for PSR B1055–52 under one-dimensional and mono-energetic approximations (Hirotani & Shibata 2002, hereafter Paper IX). We adopt $j_{\mathrm{gap}} = 0.01$ and $j_{\mathrm{out}} = 0$. The solid, dashed, dash-dotted, and dotted curves in the left panel correspond to $j_{\mathrm{in}} = 0$, 0.25, 0.5, and 0.585, respectively. It follows that the gap is located near the null surface if there is no particle injection but shifts outwards as j_{in} increases. The right panel represents $n_+(s)$ (thick solid curve), $n_-(s)$ (thin solid), and $j_{\mathrm{tot}} = j_{\mathrm{gap}} = n_+(s) + n_-(s)$ (dashed), for $j_{\mathrm{in}} = j_{\mathrm{out}} = 0$ and $j_{\mathrm{gap}} = 0.01$, which corresponds to the solution represented by the solid curve in the left panel. It follows that the pair creation mainly takes place in the inner region of the gap. This is because most of the pairs are created by the (more or less head-on) collisions between inward-directed γ-rays and surface X-rays. Since created pairs at s is proportional to the number of γ-rays emitted at larger s (via curvature process), n_- (thin curve) increases roughly exponentially with decreasing s. Note that the current one-dimensional treatment enhances pair creation, because γ-rays do not escape into the convex side due to the field-line curvature.

3.5. Mono-energetic Approximation: Gap Position vs. Injected Current

In this section, we analytically investigate why the gap position shifts outwards with increasing particle injection across the inner boundary, adopting the mono-energetic approximation.

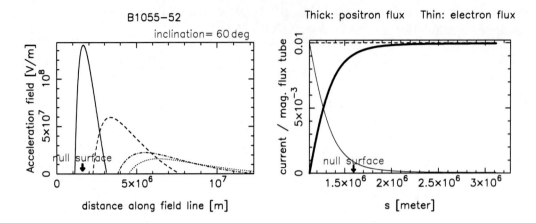

Figure 5. Solution obtained for the parameter of PSR B1055–52 with $j_{\mathrm{gap}} = 0.01$. *Left*: $E_{\parallel}(s)$ for four different injection currents: $(j_{\mathrm{in}}, j_{\mathrm{out}}) = (0,0)$ (solid), $(0.25,0)$ (dashed), $(0.5,0)$ (dash-dotted), and $(0.585,0)$ (dotted). *Right*: $n_{+}(s)$ (thick) and $n_{-}(s)$ (thin) for $(j_{\mathrm{in}}, j_{\mathrm{out}}) = (0,0)$. From Paper IX.

3.5.1.   Particle Continuity Equations

Integrating equation (9) over the momentum space, and assuming that N vanishes rapidly enough at $p_i \to \pm\infty$ ($i = 1, 2, 3$), we obtain

$$\frac{\partial \tilde{N}}{\partial t} + \vec{\nabla} \cdot \left(\langle \vec{v} \rangle \tilde{N} \right) = \tilde{S}(t, \vec{x}), \tag{27}$$

where the particle number density $\tilde{N}$ and the averaged particle velocity $\langle \vec{v} \rangle$ are defined by

$$\tilde{N}(t, \vec{x}) \equiv \int_{-\infty}^{\infty} N(t, \vec{x}, \vec{p}) d^3\vec{p}, \quad \langle \vec{v} \rangle \equiv \frac{\displaystyle\int_{-\infty}^{\infty} \vec{v} N d^3\vec{p}}{\displaystyle\int_{-\infty}^{\infty} N d^3\vec{p}}. \tag{28}$$

Since the IC scatterings and the synchro-curvature process conserve the particle number,

$$\tilde{S}(t, \vec{x}) \equiv \int_{-\infty}^{\infty} S(t, \vec{x}, \vec{p}) d^3\vec{p} \tag{29}$$

consists of pair creation and annihilation terms. For a typical pulsar magnetosphere, annihilation is negligibly small compared with creation. Therefore, we obtain

$$\tilde{S}(t, \vec{x}) = \frac{1}{c} \int_0^{\infty} dE_\gamma \left[\eta_{\mathrm{p}}(\vec{x}, E_\gamma, \mu_+) G_+ + \eta_{\mathrm{p}}(\vec{x}, E_\gamma, \mu_-) G_- \right], \tag{30}$$

where $G_+(t, \vec{x}, E_\gamma)$ and $G_-(t, \vec{x}, E_\gamma)$ designate the distribution functions of outward- and inward-directed γ-ray photons, respectively. The pair-creation redistribution functions are defined by

$$\eta_{\mathrm{p}}(\vec{x}; E_\gamma, \mu) = c \int_{-1}^{1} d\mu (1 - \mu) \int_{E_{\mathrm{th}}}^{\infty} dE_{\mathrm{x}} \frac{dN_{\mathrm{x}}}{dE_{\mathrm{x}} d\mu} \sigma_{\mathrm{p}}, \tag{31}$$

where $\sigma_{\rm p}(E_\gamma, E_{\rm x}, \mu)$ represents the pair-creation cross section, and

$$E_{\rm th} \equiv \frac{2}{1-\mu} \frac{(m_e c^2)^2}{E_\gamma}; \tag{32}$$

$\cos^{-1}\mu_+$ (or $\cos^{-1}\mu_-$) is the collision angle between the X-rays and the outwardly (or inwardly) propagating γ-rays.

Imposing a stationary condition (10), and utilizing $\nabla \cdot \boldsymbol{B} = 0$, we obtain from equation (27),

$$\pm B \frac{\partial}{\partial s} \left(\frac{\tilde{N}_\pm}{B} \right) = \frac{1}{\lambda_{\rm p}} \int_0^\infty dE_\gamma \left(G_+ + G_- \right), \tag{33}$$

The pair-creation mean free path $\lambda_{\rm p}(s)$ is defined by

$$\frac{1}{\lambda_{\rm p}} \equiv \frac{\displaystyle\int_0^\infty [\eta_{\rm p}(s, E_\gamma, \mu_+)G_+ + \eta_{\rm p}(s, E_\gamma, \mu_-)G_-]\, dE_\gamma}{\displaystyle c \int_0^\infty (G_+(s, E_\gamma) + G_-(s, E_\gamma))\, dE_\gamma}. \tag{34}$$

Since the number of created positrons is always equal to that of created electrons, the right-hand side of equation (33) is common for $\tilde{N}_+$ and $\tilde{N}_-$.

In terms of $G_\pm \equiv (\Omega B_*/2\pi ce)g_\pm$ and $N_\pm \equiv (\Omega B/2\pi ce)n_\pm$, equation (16) gives

$$\pm c \frac{\partial G_\pm}{\partial s} \approx \int_1^\infty \eta_{\rm SC} N_\pm d\Gamma, \tag{35}$$

where absorption and ICS are neglected in $S_{\gamma,\pm}$. Integrating this equation over E_γ, combining with equations (33), and assuming $\partial_s(\lambda_{\rm p} \cos \Phi) = 0$, we obtain

$$\pm \frac{d^2}{ds^2} \left(\frac{N_\pm}{B} \right) = \frac{1}{\lambda_{\rm p}c} \frac{N_+ - N_-}{B} \int_0^\infty \eta_{\rm SC} dE_\gamma. \tag{36}$$

3.5.2.   Real Charge Density in the Gap

One combination of the two independent equations (36) yields the current conservation law; that is, the total current density per magnetic flux tube,

$$j_{\rm tot} = \frac{2\pi ce}{\Omega} \frac{\tilde{N}_+(s) + \tilde{N}_-(s)}{B(s)} \tag{37}$$

is conserved along the field lines. (Note that it can be derived directly from eq. [33].) Another combination gives

$$\frac{d^2}{ds^2} \left(\frac{\tilde{N}_+ - \tilde{N}_-}{B} \right) = \frac{2}{W_\parallel} \cdot \frac{N_\gamma}{\lambda_{\rm p}} \frac{\tilde{N}_+ - \tilde{N}_-}{B}, \tag{38}$$

where

$$N_\gamma \equiv \frac{W}{c} \int_0^\infty \eta_{\rm SC}(s, \Gamma, E_\gamma) dE_\gamma \tag{39}$$

refers to the expectation value of the number of γ-rays emitted by a single particle that runs the gap width, $W_\parallel \equiv s_{\text{out}} - s_{\text{in}}$. Lorentz factor appearing in η_{SC} should be evaluated at each position s.

Exactly speaking, λ_p depends on G_+ and G_-; thus, the γ-ray distribution functions are not eliminated in equation (36). Nevertheless, for analytic (and qualitative) discussion of the gap position, we may ignore such details and adopt equation (38).

A typical γ-ray propagates the length $W_\parallel/2$ within the gap that is transversely thick. Thus, so that a stationary pair-creation cascade may be maintained, the optical depth, $W_\parallel/(2\lambda_p)$, must equal the expectation value for a γ-ray to materialize with the gap, N_γ^{-1}. We thus obtain the following condition: $W_\parallel/2 = \lambda_p/N_\gamma$. This relation holds for a self-sustaining gap in which all the particles are supplied by the pair creation. If there is an external particle injection, the injected particles also contribute for the γ-ray emission. As a result, a stationary gap can be maintain with a smaller width compared to the case of no particle injection. Taking account of such injected particles, we can constrain the half gap width as

$$\frac{W_\parallel}{2} = \frac{\lambda_p}{N_\gamma} \cdot \frac{j_{\text{gap}}}{j_{\text{tot}}}, \tag{40}$$

where j_{gap} and j_{tot} refer to the created and total current densities per unit magnetic flux tube. Equation (40) is automatically satisfied if we solve the set of Maxwell and stationary Boltzmann equations. Here, j_{gap} is related with the particle injection rate across the boundaries as follows:

$$\frac{\Omega}{2\pi ce} j_{\text{gap}} = \frac{\tilde{N}_+(s_{\text{out}})}{B(s_{\text{out}})} - \frac{\tilde{N}_+(s_{\text{in}})}{B(s_{\text{in}})} = \frac{\tilde{N}_-(s_{\text{in}})}{B(s_{\text{in}})} - \frac{\tilde{N}_-(s_{\text{out}})}{B(s_{\text{out}})}. \tag{41}$$

With the aid of identity (40), we can rewrite equation (38) into the form

$$\frac{d^2}{ds^2}\left(\frac{\tilde{N}_+ - \tilde{N}_-}{B}\right) = \frac{j_{\text{gap}}}{j_{\text{tot}}} \frac{4}{W_\parallel^2} \frac{\tilde{N}_+ - \tilde{N}_-}{B}. \tag{42}$$

To solve this differential equation, we impose the following two boundary conditions:

$$ce\frac{\tilde{N}_+(s_{\text{in}})}{B(s_{\text{in}})} = \frac{\Omega}{2\pi} j_{\text{in}}, \qquad ce\frac{\tilde{N}_-(s_{\text{out}})}{B(s_{\text{out}})} = \frac{\Omega}{2\pi} j_{\text{out}}. \tag{43}$$

Then, equations (41), (43) give

$$\frac{\tilde{N}_+ - \tilde{N}_-}{B} = -\frac{\Omega}{2\pi ce}(j_{\text{gap}} - j_{\text{in}} + j_{\text{out}}) \tag{44}$$

at $s = s_{\text{in}}$, and

$$\frac{\tilde{N}_+ - \tilde{N}_-}{B} = \frac{\Omega}{2\pi ce}(j_{\text{gap}} + j_{\text{in}} - j_{\text{out}}) \tag{45}$$

at $s = s_{\text{out}}$. Under boundary conditions (44) and (45), equation (42) is solved as

$$\frac{\tilde{N}_+ - \tilde{N}_-}{B} = \frac{\Omega}{2\pi ce}\left[j_{\text{gap}} \frac{\sinh\left(\sqrt{\frac{j_{\text{gap}}}{j_{\text{tot}}}} \frac{s - s_{\text{cnt}}}{W_\parallel/2}\right)}{\sinh\left(\sqrt{\frac{j_{\text{gap}}}{j_{\text{tot}}}}\right)} + (j_{\text{in}} - j_{\text{out}})\frac{\cosh\left(\sqrt{\frac{j_{\text{gap}}}{j_{\text{tot}}}} \frac{s - s_{\text{cnt}}}{W_\parallel/2}\right)}{\cosh\left(\sqrt{\frac{j_{\text{gap}}}{j_{\text{tot}}}}\right)}\right], \tag{46}$$

where the gap center position is defined by

$$s_{\rm cnt} \equiv \frac{s_{\rm in} + s_{\rm out}}{2}. \tag{47}$$

Note that $e(\tilde{N}_+ - \tilde{N}_-) = \rho$ represents the real charge density, which appears in the Poisson equation. Thus, substituting equation (46) into (1), we obtain

$$-\nabla^2 \Psi = \frac{2B\Omega}{c}\left[j_{\rm gap} f_{\rm odd}\left(\frac{s - s_{\rm cnt}}{W_\parallel/2}\right) + (j_{\rm in} - j_{\rm out}) f_{\rm even}\left(\frac{s - s_{\rm cnt}}{W_\parallel/2}\right) + \frac{B_\zeta}{B}\right], \tag{48}$$

where

$$f_{\rm odd}(x) \equiv \frac{\sinh\left(x\sqrt{j_{\rm gap}/j_{\rm tot}}\right)}{\sinh\left(\sqrt{j_{\rm gap}/j_{\rm tot}}\right)}, \qquad f_{\rm even}(x) \equiv \frac{\cosh\left(x\sqrt{j_{\rm gap}/j_{\rm tot}}\right)}{\cosh\left(\sqrt{j_{\rm gap}/j_{\rm tot}}\right)}. \tag{49}$$

At the inner boundary, $s = s_{\rm in}$, $(s - s_{\rm cnt})/(W_\parallel/2) = -1$ holds; therefore, we obtain

$$-\nabla^2 \Psi \approx \frac{\partial E_\parallel}{\partial s} = \frac{2B\Omega}{c}\left(-j_{\rm gap} + j_{\rm in} - j_{\rm out} + \frac{B_\zeta}{B}\right). \tag{50}$$

In the CHR picture, it is assumed that there is no particle injection across either of the boundaries (i.e., $j_{\rm in} = j_{\rm out} = 0$) and that the current density associated with the created particles becomes of the order of the typical Goldreich-Julian value (i.e., $j_{\rm gap} \sim 1$). However, it results in a negative $E_\parallel$ in the vicinity of the inner boundary, and hence in the reversal of $E_\parallel$ sign. In another word, for a transversely thin gap to have a positive $E_\parallel$ in the entire region, the right-hand side of equation (48) should be negative in almost entire region but should be *positive* in the vicinity of the inner boundary. It follows that the inner boundary of an 'outer gap' should be located close to the star, where $B_\zeta \sim B$ holds. This conclusion is consistent with what was obtained at the end of § 2..

3.5.3. Gap Position vs. Particle Injection

To examine the Poisson equation (48) analytically, we assume that the transfield thickness of the gap is greater than $W_\parallel$ and replace $\nabla^2 \Psi$ with $d^2\Psi/ds^2$. Furthermore, we neglect the current created in the gap and put $j_{\rm gap} \sim 0$.

First, consider the case when particles are injected across neither of the boundaries(i.e., $j_{\rm in} = j_{\rm out} = 0$). It follows that the derivative of the $E_\parallel$ vanishes at the null surface, where B_ζ vanishes. We may notice that $-d^2\Psi/ds^2 = dE_\parallel/ds$ is positive at the inner part of the gap and becomes negative at the outer part. The acceleration field is screened out at the boundaries by virtue of the spatial distribution of the local Goldreich-Julian charge density, $\rho_{\rm GJ}$. Therefore, we can conclude that the gap is located (or centers) around the null surface, if there is no particle injection from outside.

Secondly, consider the case when particles are injected across the inner boundary at $s = s_{\rm in}$ (or in general, when $j_{\rm in} - j_{\rm out} > 0$ holds). Since the function $f_{\rm even}$ is positive at arbitrary s, the gap center is located at a place where B_ζ is negative, that is, outside of the null surface. In particular, when $j_{\rm in} - j_{\rm out} \sim 1$ holds, $dE_\parallel/ds$ vanishes at the place where $B_\zeta \sim -B$ holds. In a vacuum, static dipole field, $B_\zeta \sim -B$ is realized along the last-open field line near to the light cylinder. Therefore, the gap should be located close to the light cylinder, if the injected particle flux across the inner boundary approaches the typical Goldreich-Julian value. We may notice here that f_{even} is less than unity, because $|s - s_{\text{cnt}}|$ does not exceed $W_\parallel/2$.

Thirdly and finally, consider the case when $j_{\text{in}} - j_{\text{out}} \sim -1$ holds. In this case, $dE_\parallel/ds$ vanishes at the place where $B_\zeta \sim B$. Therefore, an 'outer' gap should be located in the polar cap, if a Goldreich-Julian particle flux is injected across the outer boundary.

3.6. Energy Dependence of Particle Distribution Functions

The analytic conclusions derived in the foregoing section can be confirmed numerically discarding the mono-energetic approximation. To this aim, we apply the one-dimensional scheme described in §§ 3.1.–3.3. to the Vela pulsar parameters (Paper X).

To compare the effects of particle injection, we present $E_\parallel(s)$ for the three cases of $j_{\text{in}} = 0$ (solid), 0.25 (dashed), and 0.50 (dash-dotted) in figure 6. The magnetic inclination is chosen to be $\alpha_i = 75°$. We adopt $j_{\text{out}} = 0$ throughout this chapter, unless its value is explicitly specified. The abscissa denotes the distance along the last-open field line normalized by ϖ_{LC}. As the solid line shows, $E_\parallel$ is located around the null surface when there is no particle injection across either of the boundaries. Moreover, $E_\parallel$ varies quadratically, because the Goldreich-Julian charge density deviates from zero nearly linearly near to the null surface. As the dashed and dash-dotted lines indicate, the gap shifts outwards as j_{in} increases. When $j_{\text{in}} = 0.5$ for instance, the gap is located on the half way between the null surface and the light cylinder. This result is consistent with what is obtained under mono-energetic approximation (Papers VII–IX).

In the right panel, we present the characteristics of partial differential equation (12) for positrons by solid lines, together with $E_\parallel(s)$ when $j_{\text{in}} = 0.25$ and $\alpha_i = 75°$ (i.e., the dashed line in the right panel). We also superpose the equilibrium Lorentz factor that would be obtained if we assumed the balance between the curvature radiation reaction and the electrostatic acceleration, as the dotted line. It follows that the particles are not saturated at the equilibrium Lorentz factor in most portions of the gap.

In the outer part of the gap where $E_\parallel$ is decreasing, characteristics begin to concentrate; as a result, the energy distribution of outwardly propagating particles forms a 'shock' in the Lorentz factor direction. However, the particle Lorentz factors do not match the equilibrium value (dotted line). For example, near the outer boundary, the particles have larger Lorentz factors compared with the equilibrium value, because the curvature cooling scale is longer than the gap width. Thus, we must discard the mono-energetic approximation that all the particles migrate at the equilibrium Lorentz factor. We instead have to solve the energy dependence of the particle distribution functions explicitly.

In figure 7, we present the energy distribution of particles at several representative points along the field line. At the inner boundary ($s = 0.184\varpi_{\text{LC}}$), particles are injected with Lorentz factors typically less than 4×10^6 as indicated by the solid line. Particles migrate along the characteristics in the phase space and gradually form a 'shock' as the dashed line (at $s = 0.205\varpi_{\text{LC}}$) indicates, and attains maximum Lorentz factor at $s = 0.228\varpi_{\text{LC}}$ as the dash-dotted line indicates. Then they begin to be decelerated gradually and escape

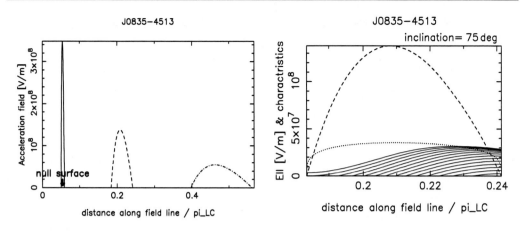

Figure 6. Results for the Vela pulsar parameter when $\alpha_i = 75°$, $j_{gap} = 4.6 \times 10^{-5}$, and $j_{out} = 0$. The abscissa designates the distance along the last-open field line normalized by the light cylinder radius. *Left*: $E_{\parallel}(s)$ for $j_{in} = 0$ (solid), 0.25 (dashed), and 0.5 (dash-dotted). *Right*: $E_{\parallel}(s)$ (dashed), the equilibrium Lorentz factor (dotted), and positronic characteristics (solid) for $j_{in} = 0.25$ (i.e., the dashed curve case in the right panel). From Paper X.

from the gap with large Lorentz factors $\sim 2.5 \times 10^7$ (dotted line) at the outer boundary, $s = s_{out} = 0.241 \varpi_{LC}$.

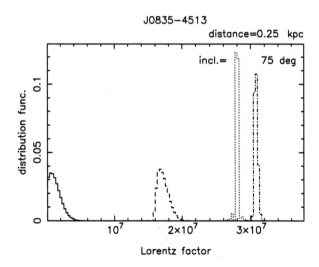

Figure 7. Particle distribution function computed for the same case as the right panel of figure 6 at four discrete positions (see text). From Paper X.

So far, we have investigate several basic features of gap electrodynamics in one-dimensional approximation. However, in a realistic pulsar magnetosphere, γ-rays propagate in the convex side to deviate from the last-open field line; thus, pair creation rate increases at higher latitudes (i.e., away from the last-open field line). In the next section, we consider the trans-field structure by solving the basic equations on the two-dimensional plane.

4.   2-Dimensional Analysis of Gap Electrodynamics

In this section, we recover θ_* dependence in equation (1), θ_* and χ dependence in equation (9), and θ_* and k^θ/k^r dependence in equation (14). We still suppress toroidal variables φ_* and k_φ. This treatment corresponds to an extension of Takata et al. (2004, hereafter TSH04; 2006, hereafter TSHC06), who solved the basic equations in the two-dimensional configuration space (see table 1), into the four-dimensional phase space (2-D configuration and 2-D momentum spaces). In § 4.1., we derive the Poisson equation that is applicable to arbitrary axisymmetric space-time and to arbitrary magnetic field configurations. We then present the Boltzmann equations for $e^\pm$'s and γ-rays, and apply the scheme to the Crab pulsar.

4.1.   Poisson Equation

Around a rotating neutron star with mass M, the background space-time geometry is given by (Lense & Thirring 1918)

$$ds^2 = g_{tt}dt^2 + 2g_{t\varphi}dtd\varphi + g_{rr}dr^2 + g_{\theta\theta}d\theta^2 + g_{\varphi\varphi}d\varphi^2, \tag{51}$$

where

$$g_{tt} \equiv \left(1 - \frac{r_{\rm g}}{r}\right)c^2, \; g_{t\varphi} \equiv ac\frac{r_{\rm g}}{r}\sin^2\theta, \tag{52}$$

$$g_{rr} \equiv -\left(1 - \frac{r_{\rm g}}{r}\right)^{-1}, \; g_{\theta\theta} \equiv -r^2, \; g_{\varphi\varphi} \equiv -r^2\sin^2\theta; \tag{53}$$

$r_{\rm g} \equiv 2GM/c^2$ indicates the Schwarzschild radius, and $a \equiv I\Omega/(Mc)$ parameterizes the stellar angular momentum. At radial coordinate r, the inertial frame is dragged at angular frequency $\omega \equiv -g_{t\varphi}/g_{\varphi\varphi} = 0.15\Omega I_{45}r_6^{-3}$, where $I_{45} \equiv I/10^{45}$ erg cm^2, and $r_6 \equiv r_*/10$ km.

Let us consider the Gauss's law,

$$\nabla_\mu F^{t\mu} = \frac{1}{\sqrt{-g}}\partial_\mu\left[\frac{\sqrt{-g}}{\rho_{\rm w}^2}g^{\mu\nu}(-g_{\varphi\varphi}F_{t\nu} + g_{t\varphi}F_{\varphi\nu})\right] = \frac{4\pi}{c^2}\rho, \tag{54}$$

where ∇ denotes the covariant derivative, the Greek indices run over t, r, θ, φ; $\sqrt{-g} = \sqrt{g_{rr}g_{\theta\theta}\rho_{\rm w}^2} = cr^2\sin\theta$ and $\rho_{\rm w}^2 \equiv g_{t\varphi}^2 - g_{tt}g_{\varphi\varphi}$, ρ the real charge density. The electromagnetic fields observed by a distant static observer are given by (Camenzind 1986a, b) $E_r = F_{rt}$, $E_\theta = F_{\theta t}$, $E_\varphi = F_{\varphi t}$, $B^r = (g_{tt} + g_{t\varphi}\Omega)F_{\theta\varphi}/\sqrt{-g}$, $B^\theta = (g_{tt} + g_{t\varphi}\Omega)F_{\varphi r}/\sqrt{-g}$, $B_\varphi = -\rho_{\rm w}^2 F_{r\theta}/\sqrt{-g}$, where $F_{\mu\nu} \equiv \partial_\mu A_\nu - \partial_\nu A_\mu$ and A_μ denotes the vector potential.

Assuming that the electromagnetic fields are unchanged in the corotating frame, we can introduce the non-corotational potential Ψ such that

$$F_{\mu t} + \Omega F_{\mu\varphi} = -\partial_\mu\Psi(r, \theta, \varphi - \Omega t), \tag{55}$$

where $\mu = t, r, \theta, \varphi$. If $F_{At} + \Omega F_{A\varphi} = 0$ $(A = r, \theta)$ holds, Ω is conserved along the field line. On the neutron-star surface, we impose $F_{\theta t} + \Omega F_{\theta\varphi} = 0$ (perfect conductor) to find that the surface is equi-potential, that is, $\partial_\theta \Psi = \partial_t \Psi + \Omega \partial_\varphi \Psi = 0$ holds. However, in a particle acceleration region, $F_{At} + \Omega F_{A\varphi}$ deviates from 0 and the magnetic field does not rigidly rotate. The deviation is expressed in terms of Ψ, which gives the strength of the acceleration electric field that is measured by a distant static observer as

$$E_\parallel \equiv \frac{\boldsymbol{B}}{B} \cdot \boldsymbol{E} = \frac{B^i}{B}(F_{it} + \Omega F_{i\varphi}) = \frac{\boldsymbol{B}}{B} \cdot (-\nabla \Psi), \tag{56}$$

where the Latin index i runs over spatial coordinates r, θ, φ.

Substituting equation (55) into (54), we obtain the Poisson equation for the non-corotational potential,

$$-\frac{c^2}{\sqrt{-g}}\partial_\mu \left(\frac{\sqrt{-g}}{\rho_w^2} g^{\mu\nu} g_{\varphi\varphi} \partial_\nu \Psi \right) = 4\pi(\rho - \rho_{GJ}), \tag{57}$$

where the general relativistic Goldreich-Julian charge density is defined as

$$\rho_{GJ} \equiv \frac{c^2}{4\pi\sqrt{-g}}\partial_\mu \left[\frac{\sqrt{-g}}{\rho_w^2} g^{\mu\nu} g_{\varphi\varphi}(\Omega - \omega)F_{\varphi\nu} \right]. \tag{58}$$

If ρ deviates from ρ_{GJ} in any region, $E_\parallel$ is exerted along $\boldsymbol{B}$. In the limit $r \gg r_g$, equation (58) reduces to the ordinary, special-relativistic expression (Goldreich and Julian 1969; Mestel 1971),

$$\rho_{GJ} \equiv -\frac{\boldsymbol{\Omega} \cdot \boldsymbol{B}}{2\pi c} + \frac{(\boldsymbol{\Omega} \times \boldsymbol{r}) \cdot (\nabla \times \boldsymbol{B})}{4\pi c}. \tag{59}$$

Instead of (r, θ, φ), it is convenient to adopt the magnetic coordinates (s, θ_*, φ_*), which was introduced in § 2.. With this coordinate system, we obtain the following form of Poisson Eq., which can be applied to arbitrary magnetic field configurations (Paper XI):

$$-\frac{c^2 g_{\varphi\varphi}}{\rho_w^2} \left(g^{ss}\partial_s^2 + g^{\theta_*\theta_*}\partial_{\theta_*}^2 + g^{\varphi_*\varphi_*}\partial_{\varphi_*}^2 \right.$$
$$\left. +2g^{s\theta_*}\partial_s\partial_{\theta_*} + 2g^{\theta_*\varphi_*}\partial_{\theta_*}\partial_{\varphi_*} + 2g^{\varphi_*s}\partial_{\varphi_*}\partial_s \right) \Psi$$
$$- \left(A^s\partial_s + A^{\theta_*}\partial_{\theta_*} + A^{\varphi_*}\partial_{\varphi_*} \right) \Psi = 4\pi(\rho - \rho_{GJ}), \tag{60}$$

$$g^{i'j'} = g^{\mu\nu}\frac{\partial x^{i'}}{\partial x^\mu}\frac{\partial x^{j'}}{\partial x^\nu} = g^{rr}\frac{\partial x^{i'}}{\partial r}\frac{\partial x^{j'}}{\partial r} + g^{\theta\theta}\frac{\partial x^{i'}}{\partial \theta}\frac{\partial x^{j'}}{\partial \theta} - \frac{k_0}{\rho_w^2}\frac{\partial x^{i'}}{\partial \varphi}\frac{\partial x^{j'}}{\partial \varphi}, \tag{61}$$

$$A^{i'} \equiv \frac{c^2}{\sqrt{-g}}\left\{ \partial_r \left[\frac{g_{\varphi\varphi}}{\rho_w^2}\sqrt{-g}g^{rr}\frac{\partial x^{i'}}{\partial r} \right] + \partial_\theta \left[\frac{g_{\varphi\varphi}}{\rho_w^2}\sqrt{-g}g^{\theta\theta}\frac{\partial x^{i'}}{\partial \theta} \right] \right\} - \frac{c^2 g_{\varphi\varphi}}{\rho_w^2}\frac{k_0}{\rho_w^2}\frac{\partial^2 x^{i'}}{\partial \varphi^2}, \tag{62}$$

where $x^1 = r$, $x^2 = \theta$, $x^3 = \varphi$, $x^{1'} = s$, $x^{2'} = \theta_*$, and $x^{3'} = \varphi_*$. The light surface, a generalization of the light cylinder, is obtained by setting $k_0 \equiv g_{tt} + 2g_{t\varphi}\Omega + g_{\varphi\varphi}\Omega^2$ to be zero (e.g., Znajek 1977; Takahashi et al., 1990). Equation (56) gives $E_\parallel = -(\partial\Psi/\partial s)_{\theta_*,\varphi_*}$. Magnetic field expansion effect is contained in $g^{\theta_*\theta_*}$, $g^{\theta_*\varphi_*}$, $g^{\varphi_*\varphi_*}$.

4.2. Particle Boltzmann Equations

In the same way as we derived equation (12), we can reduce the particle Boltzmann equations along each magnetic field line (i.e., for a constant θ_*) as

$$c \cos\chi \frac{\partial n_\pm}{\partial s} + \frac{dp}{dt}\frac{\partial n_\pm}{\partial p} + \frac{d\chi}{dt}\frac{\partial n_\pm}{\partial \chi} = S_\pm, \tag{63}$$

where the upper and lower signs correspond to the positrons (with charge $q = +e$) and electrons ($q = -e$), respectively; $p \equiv |\boldsymbol{p}|$ and

$$\frac{dp}{dt} \equiv qE_\parallel \cos\chi - \frac{P_{SC}}{c} \tag{64}$$

$$\frac{d\chi}{dt} \equiv -\frac{qE_\parallel \sin\chi}{p} + c\frac{\partial(\ln B^{1/2})}{\partial s}\sin\chi, \tag{65}$$

$$\frac{ds}{dt} = c\cos\chi. \tag{66}$$

For outward- (or inward-) migrating particles, $\cos\chi > 0$ (or $\cos\chi < 0$). Since we consider relativistic particles, we obtain $\Gamma = p/(m_e c)$. The second term in the right-hand side of equation (65) shows that the particle's pitch angle evolves due to the the variation of B (e.g., § 12.6 of Jackson 1962). For example, without $E_\parallel$, inward-migrating particles would be reflected by the magnetic mirrors. Using $n_\pm$, we can express the leptonic charge density ρ_e as

$$\rho_e = \frac{\Omega B}{2\pi c} \iint [n_+(s, \theta_*, \varphi_*, \Gamma, \chi) - n_-(s, \theta_*, \varphi_*, \Gamma, \chi)]\, d\Gamma d\chi. \tag{67}$$

The radiation-reaction force due to synchro-curvature radiation is given by (Cheng & Zhang 1996; ZC97),

$$\frac{P_{SC}}{c} = \frac{e^2\Gamma^4 Q_2}{12 r_c}\left(1 + \frac{7}{r_c^2 Q_2^2}\right), \tag{68}$$

where

$$r_c \equiv \frac{c^2}{(r_B + \rho_c)(c\cos\chi/\rho_c)^2 + r_B\omega_B^2}, \tag{69}$$

$$Q_2^2 \equiv \frac{1}{r_B}\left(\frac{r_B^2 + \rho_c r_B - 3\rho_c^2}{\rho_c^3}\cos^4\chi + \frac{3}{\rho_c}\cos^2\chi + \frac{1}{r_B}\sin^4\chi\right), \tag{70}$$

$$r_B \equiv \frac{\Gamma m_e c^2 \sin\chi}{eB}, \qquad \omega_B \equiv \frac{eB}{\Gamma m_e c} \tag{71}$$

and ρ_c is the curvature radius of the magnetic field line. In the limit of $\chi \to 0$ (or $\rho_c \to \infty$), equation (68) becomes the expression of pure curvature (or pure synchrotron) emission.

We briefly comment on the inclusion of the radiation-reaction force, P_{SC}/c, in equation (64). Except for the vicinity of the star, the magnetic field is much less than the critical value ($B_{cr} \equiv 4.41 \times 10^{13}$ G) so that quantum effects can be neglected in synchrotron radiation. Thus, we regard the radiation-reaction force, which is continuous, as an external force acting on a particle. Near the star, if $\Gamma(B/B_{cr})\sin\chi > 0.1$ holds, the energy loss rate decreases from the classical formula (Erber et al., 1966). If $\Gamma(B/B_{\mathrm{cr}})\sin\chi > 1$ holds very close to the star, the particle motion perpendicular to the field is quantized and the emission is described by the transitions between Landau states; thus, equation (64) and (68) breaks down. In this case, we artificially put $\chi = 10^{-20}$, which guarantees pure-curvature radiation after the particles have fallen onto the ground-state Landau level, avoiding to discuss the detailed quantum effects in the strong-B region. This treatment will not affect the main conclusions of this chapter, because the gap electrodynamics is governed by the pair creation taking place not very close to the star.

Let us describe some details of the collision terms $S_{\pm}$, which are given by equation (11). If we multiply $d\Gamma$ on both sides of equation (11), the first (or the second) term in the right-hand side represents the rate of particles disappearing from (or appearing into) the energy interval $m_{\mathrm{e}}c^2\Gamma$ and $m_{\mathrm{e}}c^2(\Gamma + d\Gamma)$ due to inverse-Compton (IC) scatterings; the third term does the rate of two-photon and one-photon pair creation processes.

The IC redistribution function $\eta_{\mathrm{IC}}^{\gamma}(E_{\gamma}, \Gamma, \mu_{\mathrm{c}})$ represents the probability that a particle with Lorentz factor Γ upscatters photons into energies between E_{γ} and $E_{\gamma} + dE_{\gamma}$ per unit time when the collision angle is $\cos^{-1}\mu_{\mathrm{c}}$. On the other hand, $\eta_{\mathrm{IC}}^{\mathrm{e}}(\Gamma_i, \Gamma, \mu_{\mathrm{c}})$ describes the probability that a particle changes Lorentz factor from Γ_i to Γ in a scattering. Thus, energy conservation gives

$$\eta_{\mathrm{IC}}^{\mathrm{e}}(\Gamma_i, \Gamma_f, \mu_{\mathrm{c}}) = \eta_{\mathrm{IC}}^{\gamma}[(\Gamma_i - \Gamma_f)m_{\mathrm{e}}c^2, \Gamma_i, \mu_{\mathrm{c}}]. \tag{72}$$

The quantity $\eta_{\mathrm{IC}}^{\gamma}$ is defined by the soft photon flux $dF_{\mathrm{s}}/dE_{\mathrm{s}}$ and the Klein-Nishina cross section σ_{KN} as follows (Paper X):

$$\eta_{\mathrm{IC}}^{\gamma}(E_{\gamma}, \Gamma, \mu_{\mathrm{c}}) = (1 - \beta\mu_{\mathrm{c}})$$
$$\times \int_{E_{\min}}^{E_{\max}} dE_{\mathrm{s}} \frac{dF_{\mathrm{s}}}{dE_{\mathrm{s}}} \int_{b_{i-1}}^{b_i} dE_{\gamma} \frac{dE_{\gamma}'}{dE_{\gamma}} \int_{-1}^{1} d\Omega_{\gamma}' \frac{d\sigma_{\mathrm{KN}}'(E_{\gamma}, \Gamma, \mu_{\mathrm{c}})}{dE_{\gamma}' d\Omega_{\gamma}'} \tag{73}$$

where $\beta \equiv \sqrt{1 - 1/\Gamma^2}$ is virtually unity, Ω_{γ} the solid angle of upscattered photon, the prime denotes the quantities in the electron (or positron) rest frame, and $E_{\gamma} = (b_{i-1} + b_i)/2$. In the rest frame of a particle, a scattering always takes place well above the resonance energy. Thus, the Klein-Nishina cross section can be applied to the present problem. The soft photon flux per unit photon energy E_{s} [$s^{-1}\mathrm{cm}^{-2}$] is written as $dF_{\mathrm{s}}/dE_{\mathrm{s}}$ and is given by the surface blackbody emission with redshift corrections at each distance from the star.

The differential pair-creation redistribution function is given by

$$\frac{\partial\eta_{\gamma\gamma}}{\partial\Gamma}(E_{\gamma}, \Gamma, \mu_{\mathrm{c}}) = (1 - \mu_{\mathrm{c}}) \int_{E_{\mathrm{th}}}^{\infty} dE_{\mathrm{s}} \frac{dF_{\mathrm{s}}}{dE_{\mathrm{s}}} \frac{d\sigma_{\mathrm{p}}(E_{\gamma}, \Gamma, \mu_{\mathrm{c}})}{d\Gamma}, \tag{74}$$

where the pair-creation threshold energy is defined by equation (32), and the differential cross section is given by

$$\frac{d\sigma_{\mathrm{p}}}{d\Gamma} = \frac{3}{8}\sigma_{\mathrm{T}} \frac{1 - \beta_{\mathrm{CM}}^2}{E_{\gamma}} \times \left[\frac{1 + \beta_{\mathrm{CM}}^2(2 - \mu_{\mathrm{CM}}^2)}{1 - \beta_{\mathrm{CM}}^2\mu_{\mathrm{CM}}^2} - \frac{2\beta_{\mathrm{CM}}^4(1 - \mu_{\mathrm{CM}}^2)^2}{(1 - \beta_{\mathrm{CM}}^2\mu_{\mathrm{CM}}^2)^2}\right]; \tag{75}$$

σ_{T} refers to the Thomson cross section and the center-of-mass quantities are defined as

$$\mu_{\mathrm{CM}} \equiv \pm\frac{2\Gamma m_{\mathrm{e}}c^2 - E_{\gamma}}{\beta_{\mathrm{CM}}E_{\gamma}}, \quad \beta_{\mathrm{CM}}^2 \equiv 1 - \frac{2(m_{\mathrm{e}}c^2)^2}{(1 - \mu_{\mathrm{c}})E_{\mathrm{s}}E_{\gamma}}. \tag{76}$$

Since a convenient form of $\partial \eta_{\gamma B}/\partial \Gamma$ is not given in the literature, we simply assume that all the particles are created at the energy $\Gamma m_e c^2 = E_\gamma/2$ for magnetic pair creation. This treatment does not affect the conclusions in this chapter.

Let us briefly mention the electric current per magnetic flux tube. With projected velocities, $c \cos \chi$, along the field lines, electric current density in units of $\Omega B/(2\pi)$ is given by

$$j_{\text{gap}}(s, \theta_*) = j_e(s, \theta_*) + j_{\text{ion}}(\theta_*), \tag{77}$$

where

$$j_e \equiv \iint (n_- + n_+) \cos \chi \, dp d\chi; \tag{78}$$

j_{ion} denotes the current density carried by the ions emitted from the stellar surface. Since dp/dt and $d\chi/dt$ in equation (63) depend on momentum variables p and χ, j_e and hence j_{gap} does not conserve along the field line in an exact sense. Nevertheless, j_{gap} is virtually kept constant for s, because the returning particles, of which pitch angles satisfy $|\cos \chi| \ll 1$, occupy only a small population at each point.

4.3. Gamma-ray Boltzmann Equations

Imposing the stationary condition (10), or equivalently, assuming that g depends on φ and t as $g = g(r, \theta, \varphi - \Omega t, \mathbf{k})$, we obtain

$$\left(c\frac{k^\varphi}{|\mathbf{k}|} - \Omega \right) \frac{\partial g}{\partial \bar{\varphi}} + c\frac{k^r}{|\mathbf{k}|} \frac{\partial g}{\partial r} + c\frac{k^\theta}{|\mathbf{k}|} \frac{\partial g}{\partial \theta} = S_\gamma(r, \theta, \bar{\varphi}, c|\mathbf{k}|, k^r, k^\theta, k_\varphi), \tag{79}$$

where $\bar{\varphi} = \varphi - \Omega t$. To compute k^i, we have to solve the photon propagation in the curved spacetime. Since the wavelength is much shorter than the typical system scales, geometrical optics gives the evolution of momentum and position of a photon by the Hamilton-Jacobi equations,

$$\frac{dk_r}{d\lambda} = -\frac{\partial k_t}{\partial r}, \quad \frac{dk_\theta}{d\lambda} = -\frac{\partial k_t}{\partial \theta} \tag{80}$$

$$\frac{dr}{d\lambda} = \frac{\partial k_t}{\partial k_r}, \quad \frac{d\theta}{d\lambda} = \frac{\partial k_t}{\partial k_\theta}, \tag{81}$$

where the parameter λ is defined so that $cd\lambda$ represents the distance (i.e., line element) along the ray path. The photon energy at infinity k_t and the azimuthal wave number $-k_\varphi$ are conserved along the photon path in a stationary and axisymmetric spacetime (e.g., in the spacetime described by eqs. [51]–[53]). Hamiltonian k_t can be expressed in terms of k_r, k_θ, k_φ, r, θ from the dispersion relation $k^\mu k_\mu = 0$, which is a quadratic equation of k_μ ($\mu = t, r, \theta, \varphi$). Thus, we have to solve the set of four ordinary differential equations (80) and (81) for the four quantities, k_r, k_θ, r, and θ along the ray. Initial conditions at the emitting point are given by $k^i/|\mathbf{k}| = \pm B^i/|\mathbf{B}|$, where $i = r, \theta, \varphi$; the upper (or lower) sign is chosen for the γ-rays emitted by an outward- (or inward-) migrating particle. When a photon is emitted with energy E_{local} by the particle of which angular velocity is $\dot{\varphi}$, it is related with k_t and $-k_\varphi$ by the redshift relation, $E_{\text{local}} = (dt/d\tau)(k_t + k_\varphi \dot{\varphi})$, where $dt/d\tau$ is solved from $(dt/d\tau)^2(g_{tt} + 2g_{t\varphi}\dot{\varphi} + g_{\varphi\varphi}\dot{\varphi}^2) = 1$. To express the energy dependence of g, we regard g as a function of $k_t = E_\gamma$ (i.e., observed photon energy).

In this chapter, in accordance with the two-dimensional analysis of equations (60) and (63), we neglect $\bar{\varphi}$ dependence of g, by ignoring the first term in the left-hand side of equation (79). In addition, we neglect the aberration of photons and simply assume that the γ-rays do not have angular momenta and put $k_\varphi = 0$. The aberration effects are important when we discuss how the outward-directed γ-rays will be observed. However, they can be correctly taken into account only when we compute the propagation of emitted photons in the three-dimensional magnetosphere. Moreover, they are not essential when we investigate the electrodynamics, because the pair creation is governed by the specific intensity of inward-directed γ-rays, which are mainly emitted in a relatively inner region of the magnetosphere. Thus, it seems reasonable to adopt $k_\varphi = 0$ when we investigate the two-dimensional gap electrodynamics.

We linearly divide the longitudinal distance into 400 grids from $s = 0$ (i.e., stellar surface) to $s = 1.4\varpi_{LC}$, and the meridional coordinate into 16 field lines from $\theta_* = \theta_*^{max}$ (i.e., the last-open field line) to $\theta_* = \theta_*^{min}$ (i.e., gap upper boundary), and consider only $\varphi_* = 0$ plane (i.e., the field lines threading the stellar surface on the plane formed by the rotation and magnetic axes). To solve the particle Boltzmann equations (63), we adopt the Cubic Interpolated Propagation (CIP) scheme with the fractional step technique to shift the profile of the distribution functions $n_\pm$ in the direction of the velocity vector in the two-dimensional momentum space. To solve the γ-ray Boltzmann equation (79), on the other hand, we do not have to compute the advection of g in the momentum space, because only the spatial derivative terms remain after integrating over γ-ray energy bins, which are logarithmically divided from $\beta_1 = 0.511$ MeV to $\beta_{29} = 28.7$ TeV into 29 bins. The γ-ray propagation directions, k^θ/k^r, are divided linearly into 180 bins every $\Delta\theta_\gamma = 2$ degrees. Since the specific intensity in ith energy bin at height $\theta_* = \theta_*^k = \theta_*^{max} - (k/16)(\theta_*^{max} - \theta_*^{min})$, is given by

$$g_{i,k,l}(s) = \frac{c}{\Delta\theta_\gamma \Delta\phi_\gamma} \int_{b_{i-1}}^{b_i} g[s, \theta_*^k, E_\gamma, (k^\theta/k^r)_l] \, dE_\gamma, \tag{82}$$

the observed γ-ray energy flux at distance d is calculated as

$$F_{i,l} = \frac{\Delta y \sum_k \Delta z_k \, g_{i,k,l}}{d^2}, \tag{83}$$

where Δy denotes the azimuthal dimension of the gap at longitudinal distance s ($= \varpi_{LC}$ in this chapter), Δz_k the meridional thickness between two field lines with $\theta_* = \theta_{*k}$ and θ_{*k+1}, and $i = 1, 2, 3, \ldots, 28$, $k = 1, 2, 3, \ldots, 15$, $l = 1, 2, 3, \ldots, 180$. To compute the phase-averaged spectrum, we set the azimuthal width of the γ-ray propagation direction, $\Delta\phi_\gamma$ to be π radian.

Equation (15) describes the γ-ray absorption and creation rate within the gap. However, to compute observable fluxes, we also have to consider the synchrotron emission by the secondary, tertiary, and higher-generation pairs that are created outside of the gap. If an electron or positron is created with energy $\Gamma_0 m_e c^2$ and pitch angle χ_0, it radiates the following number of γ-rays (in units of $\Omega B_*/2\pi ce$) in energies between b_{i-1} and b_i:

$$\frac{dg_i}{dn} = \frac{2\pi ce}{\Omega B_*} \int_0^\infty dt \int_{b_{i-1}}^{b_i} \frac{1}{E_\gamma} \frac{dW}{dt dE_\gamma} dE_\gamma, \tag{84}$$

where

$$\frac{dW}{dtdE_\gamma} = \frac{\sqrt{3}e^3 B \sin\chi_0}{hm_e c^2} F\left(\frac{E_\gamma}{E_c}\right), \quad F(x) \equiv x \int_x^\infty K_{5/3}(\xi)d\xi, \tag{85}$$

$$m_e c^2 \frac{d\Gamma}{dt} = -\frac{2}{3}\frac{e^4 B^2 \sin^2\chi_0}{m_e^2 c^3}\Gamma^2; \tag{86}$$

$K_{5/3}$ is the modified Bessel function of $5/3$ order, and $E_c \equiv (3h/4\pi)(eB\Gamma^2 \sin\chi_i)/(m_e c)$ is the synchrotron critical energy at Lorentz factor Γ. Substituting equations (85) and (86) into (84), we obtain

$$\frac{dg_i}{dn} = \frac{2\pi ce}{\Omega B_*}\frac{3\sqrt{3}m_e^2 c^3}{2heB\sin\chi_0}\int_1^{\Gamma_0}\frac{d\Gamma}{\Gamma^2}\int_{b_{i-1}/E_c}^{b_i/E_c} dy \int_y^\infty K_{5/3}(\xi)d\xi. \tag{87}$$

Note that we assume that particle pitch angle is fixed at $\chi = \chi_0$, because ultra-relativistic particles emit radiation mostly in the instantaneous velocity direction, preventing pitch-angle evolution. Once particles lose sufficient energies, they preferentially lose perpendicular momentum; nevertheless, such less-energetic particles hardly emit synchrotron photons above MeV energies. On these grounds, to incorporate the synchrotron radiation of higher-generation pairs created outside of the gap, we add $\int_1^\infty (dn/d\Gamma_0)(dg_i/dn)d\Gamma_0$ to compute the emission of γ-rays in the energy interval $[b_{i-1}, b_i]$ in the right-hand side of equation (14), where $dn/d\Gamma_0$ denotes the particles created between position s and $s+ds$ in Lorentz factor interval $[\Gamma_0, \Gamma_0 + d\Gamma_0]$.

4.4. Boundary Conditions

In order to solve the set of partial differential equations (60), (63), and (79) for Ψ, $n_\pm$, and g, we must impose appropriate boundary conditions. We assume that the gap *lower* boundary (fig. 2), $\theta_* = \theta_*^{\max}$, coincides with the last open field line, which is defined by the condition that $\sin\theta\sqrt{-g_{rr}}B^r + \cos\theta\sqrt{-g_{\theta\theta}}B^\theta = 0$ is satisfied at the light cylinder on the surface $\varphi_* = 0$. Moreover, we assume that the *upper* boundary coincides with a specific magnetic field line and parameterize this field line with $\theta_* = \theta_*^{\min}$. In general, $\theta_*^{\min}$ is a function of φ_*; however, we consider only $\varphi_* = 0$ in this chapter. Determining the upper boundary from physical consideration is a subtle issue, which is beyond the scope of this chapter. Therefore, we treat $\theta_*^{\min}$ as a free parameter. We measure the trans-field thickness of the gap with

$$h_m \equiv \frac{\theta_*^{\max} - \theta_*^{\min}}{\theta_*^{\max}}. \tag{88}$$

If $h_m = 1.0$, it means that the gap exists along all the open field lines. On the other hand, if $h_m \ll 1$, the gap becomes transversely thin and θ_* derivatives dominate in equation (60). To describe the trans-field structure, we introduce the fractional height as

$$h \equiv \frac{\theta_*^{\max} - \theta_*}{\theta_*^{\max}}. \tag{89}$$

Thus, the lower and upper boundaries are given by $h = 0$ and $h = h_m$, respectively.

The *inner* boundary is assumed to be located at the stellar surface. For the *outer* boundary, we solve the Poisson equation to a large enough distance, $s = 1.4\varpi_{LC}$, which is located outside of the light cylinder. This mathematical outer boundary is introduced only for convenience in order that the $E_\parallel$ distribution inside of the light cylinder may not be influenced by the artificially chosen outer boundary position when we solve the Poisson equation. Since the structure of the outer-most part of the magnetosphere is highly unknown, we artificially set $E_\parallel = 0$ if the distance from the rotation axis, ϖ, becomes greater than $0.90\varpi_{LC}$. Under this artificially suppressed $E_\parallel$ distribution in $\varpi > 0.90\varpi_{LC}$, we solve the Boltzmann equations for outward-propagating particles and γ-rays in $0 < s < 1.4\varpi_{LC}$. For inward-propagating particles and γ-rays, we solve only in $0 < s < 0.90\varpi_{LC}$. The position of the mathematical outer boundary ($1.4\varpi_{LC}$ in this case), little affects the results by virtue of the artificial boundary condition, $E_\parallel = 0$ for $\varpi > 0.9\varpi_{LC}$. On the other hand, the artificial outer boundary condition, $E_\parallel = 0$ for $\varpi > 0.9\varpi_{LC}$, affects the calculation of outward-directed γ-rays to some degree; nevertheless, it little affects the electrodynamics in the inner part of the gap ($s < 0.5\varpi_{LC}$), which is governed by the absorption of inward-directed γ-rays.

First, to solve the elliptic-type equation (60), we impose $\Psi = 0$ on the lower, upper, and inner boundaries. At the mathematical outer boundary ($s = 1.4\varpi_{LC}$), we impose $\partial\Psi/\partial s = 0$. Generally speaking, the solved $E_\parallel = -(\partial\Psi/\partial s)_{s\to 0}$ under these boundary conditions does not vanish at the stellar surface. Let us consider how to cancel this remaining electric field.

For a super-GJ current density in the sense that $\rho_e - \rho_{GJ} < 0$ holds at the stellar surface, equation (60) gives a positive electric field near the star. In this case, we assume that ions are emitted from the stellar surface so that the additional positive charge in the thin non-relativistic region may bring $E_\parallel$ to zero (for the possibility of free ejection of ions due to a low work function, see Jones 1985, Neuhauser et al., 1986, 1987). The column density in the non-relativistic region becomes (SAF78)

$$\Sigma_{NR} = \frac{1}{2\pi}\sqrt{\frac{c\Omega B_*}{q/m}}\,j_{ion}, \tag{90}$$

where q/m represents the charge-to-mass ratio of the ions and j_{ion} the ionic current density in units of $\Omega B_*/(2\pi)$. Equating $4\pi\Sigma_{NR}$ to $-(\partial\Psi/\partial s)_{s\to 0}$ calculated from relativistic positrons, electrons and ions, we obtain the ion injection rate j_{ion} that cancels $E_\parallel$ at the stellar surface.

For a sub-GJ current density in the sense that $\rho_e - \rho_{GJ} > 0$ holds at the stellar surface, Ψ increases outwards near the star to peak around $s = 0.02\varpi_{LC} \sim 0.10\varpi_{LC}$, depending on α_i and $\rho_e(s = 0)$, then decrease to become negative in the outer magnetosphere. That is, $-(\partial\Psi/\partial s)_{s\to 0} < 0$ holds in the inner region of the gap. In this case, we assume that electrons are emitted from the stellar surface and fill out the region where $\Psi > 0$; thus, we artificially put $\Psi = 0$ if $\Psi > 0$ appears. Even though a non-vanishing, positive $E_\parallel$ is remained at the inner boundary, which is located away from the stellar surface, we neglect such details. This is because the gap with a sub-GJ current density is found to be inactive and hence less important, as will be demonstrated in the next section.

Secondly, to solve the hyperbolic-type equations (63) and (79), we assume that neither positrons nor γ-rays are injected across the inner boundary; thus, we impose

$$n_+(s_{in}, \theta_*, \Gamma, \chi) = 0, \quad g(s_{in}, \theta_*, E_\gamma, \theta_\gamma) = 0 \tag{91}$$

for arbitrary θ_*, Γ, $0 < \chi < \pi/2$, E_γ, and $\cos(\theta_\gamma - \theta_B) > 0$, where θ_B designates the outward magnetic field direction. In the same manner, at the outer boundary, we impose

$$n_-(s_{out}, \theta_*, \Gamma, \chi) = 0, \quad g(s_{out}, \theta_*, E_\gamma, \theta_\gamma) = 0 \tag{92}$$

for arbitrary θ_*, Γ, $\pi/2 < \chi < \pi$, E_γ, and $\cos(\theta_\gamma - \theta_B) < 0$.

4.5. Application to the Crab Pulsar

We next apply the scheme to the Crab pulsar, adopting four free parameters, α_i, μ, kT_s, and h_m. Other quantities such as gap geometry on the poloidal plane, exerted $E_\parallel$ and potential drop, particle density and energy distribution, the γ-ray flux and spectrum, as well as the created pairs outside of the gap, are uniquely determined if we specify these four parameters.

The Crab pulsar has been studied from radio, optical, X-ray to γ-ray wavelength since its discovery (Staelin & Reifenstein 1968; Comella et al., 1969). Its period and period derivative are $P = 33.0$ ms and $\dot{P} = 4.20 \times 10^{-13}$ s s^{-1}. Using magnetic dipole radiation formula for an orthogonal rotator, the observed spin down luminosity 4.46×10^{38} ergs s^{-1} gives $\mu = 3.80 \times 10^{30}(d/2\text{kpc})^2$. From soft X-ray observations, 180 eV is obtained (Tennant et al., 2001) as the upper limit of the cooling neutron-star blackbody temperature, kT.

4.5.1. Sub-GJ Solution

Adopting $kT = 100$ eV, $\mu = 4.0 \times 10^{30}$ G cm^3, and the magnetic inclination $\alpha_i = 70°$, which is more or less close to the value (65°) suggested by a three-dimensional analysis in the traditional outer gap model, we obtain a nearly vacuum solution (fig. 8) for a geometrically thin case $h_m = 0.047$. In the left panel, we present $E_\parallel(s, h)$ at five discrete heights. The solutions become similar to the vacuum one obtained in CHR86a,b. For one thing, the inner boundary is located slightly inside of the null surface. What is more, $E_\parallel$ maximizes at the central height, $h = h_m/2$, and remains roughly constant in the entire region of the gap. The solved $E_\parallel$ distributes almost symmetrically with respect to the central height; for example, the dashed and dash-dotted curves nearly overlap each other. The gap has no outer termination within the light cylinder. Since the inner boundary coincides with the place where Ψ vanishes, the region between the star and the inner boundary has $\Psi > 0$. Therefore, negative charges pulled from the stellar surface with $E_\parallel \equiv -\partial_s \Psi < 0$ populate only inside of the inner boundary, in which $\Psi < 0$ holds. Similar solutions are obtained for a thiner gap, $h_m < 0.047$, even though there appears a small $E_\parallel$ peak near the null surface, which is less important.

CHR86a,b first considered this kind of vacuum solutions and suggested two outer gaps can be formed for each magnetic pole Considering a fan beam (instead of a pencil beam in the inner-gap model or a funnel beam in the inner-slot-gap model) as the emission morphology, they discussed the formation of cusped photon peak. Extending this morphological emission model, Stanford group (RY95) discussed the observed properties of individual γ-ray pulsars, their radio to γ-ray pulse offsets, and the radio- vs. γ-ray detection probabilities. Assuming $E_\parallel \propto s^{-1}$ and a power-law energy distribution accelerated $e^\pm$'s, R96 estimated the evolution of high-energy flux efficiencies and beaming fractions to discuss

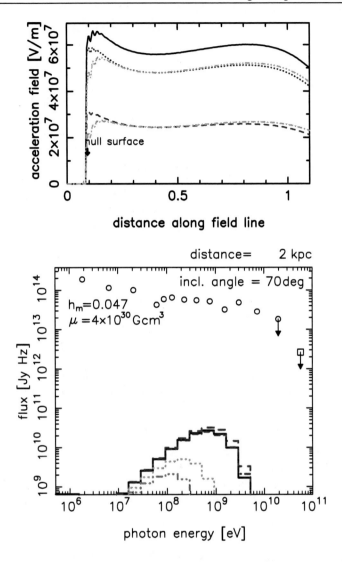

Figure 8. Traditional outer-gap solution obtained for the Crab pulsar with $\alpha_i = 70°$ and $h_m = 0.047$. *Top*: The field-aligned electric field at discrete heights h ranging from $2h_m/16$, $5h_m/16$, $8h_m/16$, $11h_m/16$, $14h_m/16$, with dashed, dotted, solid, dash-dot-dot-dot, and dash-dotted curves, respectively. The abscissa indicates the distance along the field line from the star in the unit of the light-cylinder radius. The null surface position at the height $h = h_m/2$ is indicated by the down arrow. *Bottom*: Calculated phase-averaged spectra of the pulsed, outward-directed γ-rays. The flux is averaged over the meridional emission angles (see Paper XI for details).

the detection statistics. Another group in Hong Kong (ZC97) examined the minimum trans-field thickness, h_m, of the gap, imposing that the γ-γ pair creation criterion is met. They estimated the soft photon field emitted from the heated polar-cap surface by the bombardment of gap-accelerated charged particles, adopting essentially the same $E_\parallel$ solution as Fig. 8. Extending their work, CRZ00 developed a three-dimensional outer magnetospheric gap model to examine the double-peak light curves with strong inter-pulse emission, and estimated phase-resolved γ-ray spectra by assuming that the charged particles are accelerated to the Lorentz factors at which the curvature radiation-reaction force balances with the electrostatic acceleration.

The outer gap models of these two groups have been successful in explaining the observed light curves, particularly in reproducing the wide separation of the two peaks, without invoking a very small inclination angle (as in inner-gap models). However, if we solve Eq. (60) self-consistently with the particle and γ-ray Boltzmann equations (Paper XI), we find that the γ-ray flux obtained for $h_{\mathrm{m}} < 0.047$ (i.e., traditional outer-gap models) is insufficient (right panel of fig. 8). Thus, we have to consider a transversely thicker gap, which exerts a larger $E_{\parallel}$ because of the less-efficient screening due to the two zero-potential walls at $h = 0$ and h_{m}. If the created current increases due to the increased h_{m}, the gap inner boundary deviates the null surface and shifts inwards to touch the stellar surface at last (Paper X).

4.5.2. Super-GJ Solution

For $h_{\mathrm{m}} > 0.047$, the created current density j_{e} (see eq. [77]) becomes super Goldreich-Julian in the sense that $\rho < \rho_{\mathrm{GJ}} < 0$ holds at the inner boundary (left panel of fig. 9). The predicted γ-ray flux is much larger than the sub-GJ case of $h_{\mathrm{m}} \leq 0.047$ (i.e., traditional outer-gap solution). Moreover, the copious pair creation leads to a substantial screening of $E_{\parallel}$ in the inner region, as the right panel shows. In this screening region, $\rho(s, h)$ distributes (fig. 10) so that $E_{\parallel}$ may virtually vanish. Because of the ion emission from the stellar surface, the total charge density ρ is given by $\rho = \rho_{\mathrm{e}} + \rho_{\mathrm{ion}}$, where ρ_{e} denotes the sum of positronic and electronic charge densities, while ρ_{ion} does the ionic one. We should notice here that even if $E_{\parallel} \approx 0$ occurs by the discharge of created pairs in most portions of the gap, the negative $\rho_{\mathrm{e}} - \rho_{\mathrm{GJ}}$ inevitably exerts a strong positive $E_{\parallel}$ at the surface (see Paper XI for details), thereby extracting ions from the surface for the solution with super-GJ current. On these grounds, we can regard this modern outer-gap solution as a mixture of the traditional inner-gap model, which extracts electrons from the surface with $E_{\parallel} < 0$, and the traditional outer-gap model, which exerts positive $E_{\parallel}$ because of the negativity of $\rho - \rho_{\mathrm{GJ}}$.

There is space here only for a brief comments on the Lorentz-factor and pitch-angle dependence of particle distribution functions, $n_{\pm}$ (see Paper XI for details). First, mono-energetic approximation is not good for positrons, which migrate outwards. This is because pairs are created at various points in the gap and follow different characteristics (see solid curves in the right panel of fig. 6), and because a small portions (a few percent) of positrons up-scatter the surface blackbody photons to lose energies. Positrons are mostly created with inward momenta initially and return by the positive $E_{\parallel}$ to lose most of their perpendicular momentum by synchrotron radiation. Thus, their radiation in the outer magnetosphere can be safely approximated by the pure curvature formula. Secondly, electrons efficiently up-scatter surface photons (by more or less head-on collisions) to have a broad energy spectra. For example, at $s = 0.4\varpi_{\mathrm{LC}}$, they broadly distribute in $10^4 < \Gamma < 10^7$ and $10^{-8} < \sin\chi < 10^{-3.5}$. It follows that the pure curvature formula completely breaks down for electrons, which migrate inwards in the inner magnetosphere ($B > 10^7$ G) and that we must adopt the synchro-curvature formula, instead of pure curvature one.

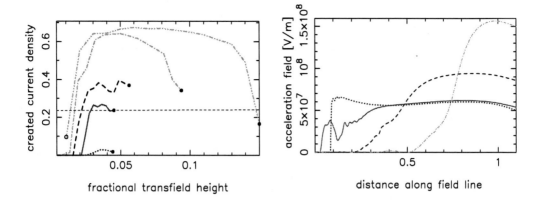

Figure 9. Modern outer-gap solution obtained for the Crab pulsar with $\alpha_i = 70°$ and $h_m \geq$ 0.048: The dotted, solid, dashed, dash-dotted, and dash-dot-dot-dotted curves corresponds to $h_m = 0.047, 0.048, 0.060, 0.100$, and 0.160, respectively. *Left*: Created current density j_e (in unit of $\Omega B/2\pi$) as a function of the transfield thickness h. If j_e appears below (or above) the dashed line, $|c\rho_{GJ}/(\Omega B/2\pi)||_{s=0}$, the solution is sub- (or super-) GJ current. *Right*: $E_{\|}(s, h_m/2)$. From Paper XI.

5. Discussion

In summary, we have quantitatively examined the stationary pair-creation cascade in an outer magnetosphere, by solving the set of Maxwell and Boltzmann equations. By one-dimensional analysis, it is revealed that the gap position shifts by the injection of charged particles across the boundary. For example, the gap is located near to the light cylinder (or the stellar surface) if the injection rate across the inner (or outer) boundary approaches the typical Goldreich-Julian value. This conclusion is, in fact, unchanged if the gap has a two-dimensional structure. It should be emphasized that the particle energy distribution is not represented by a power law, as assumed in some of previous outer-gap models. Moreover, applying the two-dimensional scheme to the Crab pulsar, we find that the solution represents the traditional outer-gap model if the created current density, j_e, is small compared to the Goldreich-Julian (GJ) value. However, in this case, the predicted γ-ray flux for the Crab pulsar is too small to explain the observed value. We find a new accelerator solution that has a super-GJ current density and extends from the stellar surface to the outer magnetosphere. This new solution possesses an acceleration electric field, $E_{\|}$, that is substantially screened in the inner part. However, the negative effective charge density, $\rho_{eff} \equiv \rho - \rho_{GJ}$ results in a non-vanishing, positive $E_{\|}$ in the inner-most region, which extracts ions from the surface. It is essential to examine the pitch-angle evolution of the created particles, because the inward-migrating particles emit γ-rays, which governs the gap electrodynamics through pair creation, via synchro-curvature process rather than pure-curvature one. The resultant spectral shape of the outward-directed γ-rays is consistent with the existing observations.

We consider the stability of such a gap in the next subsection. We then compare the new gap solution with an existing model.

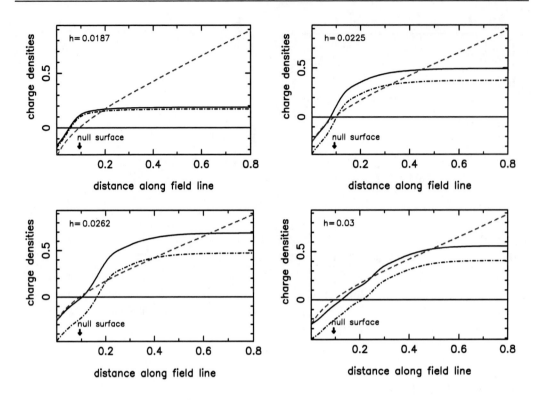

Figure 10. Modern outer-gap solution of the total (solid), created (dash-dotted), and Goldreich-Julian (dashed) charge densities in $\Omega B(s, h)/(2\pi c)$ unit, for $\alpha_i = 70°$ and $h_m = 0.060$ at four transfield heights, h. Because of an ion emission from the stellar surface, the total charge density deviates from the created one. From Paper XI.

5.1. Stability of the Gap

Let us discuss the electrodynamic stability of the gap, by considering whether an initial perturbation of some quantity tends to be canceled or not. In this chapter, we consider that the soft photon field is given and unchanged when gap quantities vary. Thus, let us first consider the case when the soft photon field is fixed. Imagine that the created pairs are decreased as an initial perturbation. It leads to an increase of the potential drop due to less efficient screening by the discharged pairs, and hence to an increase of particle energies. Then the particles emit synchro-curvature radiation efficiently, resulting in an increase of the created pairs, which tends to cancel the initial decrease of created pairs.

Let us next consider the case when the soft photon field also changes. Imagine again that the created pairs are decreased as an initial perturbation. It leads to an increase of particle energies in the same manner as discussed just above. The increased particle energies increase not only the number and density of synchro-curvature γ-rays, but also the surface blackbody emission from heated polar caps and the secondary magnetospheric X-rays. Even though neither the heated polar-cap emission nor the magnetospheric emission are taken into account as the soft photon field illuminating the gap in this chapter, they all work, in general, to increase the pair creation within the gap, which cancels the initial decrease of created pairs more strongly than the case of the fixed soft photon field.

Because of such negative feedback effects, solution exists in a wide parameter space. For example, the created current density is almost unchanged for a wide range of h_m (e.g., compare the dash-dotted and dash-dot-dot-dotted curves in the left panel of fig. 9). On these grounds, although the perturbation equations are not solved under appropriate boundary conditions for the perturbed quantities, we conjecture that the particle accelerator is electrodynamically stable, irrespective whether the X-ray field illuminating the gap is thermal or non-thermal origin.

5.2. Comparison with Polar-Slot Gap Model

It is worth comparing the present results with the polar-slot-gap model proposed by Muslimov and Harding (2003; 2004a,b, hereafter MH04a,b), who obtained a quite different solution (e.g., negative $E_\parallel$ in the gap) solving essentially the same equations under analogous boundary conditions for the same pulsar as in the present work. The only difference is the transfield thickness of the gap (i.e., h_m). Estimating the transfield thickness to be $\Delta l_{SG} \sim h_m r_* \sqrt{r/\varpi_{LC}}$, which is a few hundred times thinner than the present work, they extended the solution (near the polar cap surface) that was obtained by MT92 into the higher altitudes (towards the light cylinder). Because of this very small Δl_{SG}, emitted γ-rays do not efficiently materialize within the gap; as a result, the created and returned positrons from the higher altitudes do not break down the original assumption of SCLF near the stellar surface.

To avoid the reversal of $E_\parallel$ in the gap (from negative near the star to positive in the outer magnetosphere), or equivalently, to avoid the reversal of the sign of the effective charge density, $\rho_{eff} = \rho - \rho_{GJ}$, along the field line, MH04a and MH04b assumed that ρ_{eff}/B nearly vanishes and remains constant above a certain altitude, $s = s_c$, where s_c is estimated to be within a few neutron star radii. Because of this assumption, $E_\parallel$ is suppress at a very small value and the pair creation becomes negligible in the entire gap. In another word, the enhanced screening is caused not only by the proximity of two conducting boundaries, but also by the assumption of $\partial(\rho_{eff}/B)/\partial s = 0$ within the gap (see eq. [5]). To justify this ρ/B distribution, MH04a and MH04b proposed an idea that ρ should grow by the *cross field motion* of charges due to the toroidal forces, and that ρ_{eff}/B is a small constant so that $c\rho_{eff}/B$ may not exceed the flux of the emitted charges from the star, which ensures the *equipotentiality* of the slot-gap boundaries (see § 2.2 of MH04a for details).

The cross-field motion becomes important if particles gain angular momenta as they migrate outwards to pick up energies which is a non-negligible fraction of the difference of the cross-field potential between the two conducting boundaries. Denoting the fraction as ϵ, we obtain $\Gamma m_e c^2 \dot{\varphi} \Omega(\varpi/c)^2 = \epsilon e B \Delta l_{SG}$ (Mestel 1985; eq. [12] of MH04a). If we substitute their estimate $\Delta l_{SG} \sim r_*/20$, we obtain $\epsilon \sim 0.33(\dot{\varphi}/\Omega)\gamma_7 B_6^{-1} r_6^{-3}(\varpi/\varpi_{LC})^2$, where $\gamma_7 = \Gamma/10^7$, $B_6 = B_*/10^6\,\mathrm{G}$, and $r_6 = r_*/10\,\mathrm{km}$; therefore, the cross-field motion becomes important in the outer magnetosphere within their (transversely very thin) slot-gap model.

As for the equipotentiality of the boundaries, it seems reasonable to suppose that $c|\rho_{eff}|/B < c|\rho_*|/B_*$ should be held at any altitudes in the gap, as MH04a suggested, where ρ_* denotes the real charge density at the stellar surface. However, the assumption that ρ_{eff}/B is a small positive constant may be too strong, because it is only a sufficient condition of $c|\rho_{\text{eff}}|/B < c|\rho_*|/B_*$.

In this chapter, on the other hand, we assume that the magnetic fluxes threading the gap is unchanged, considering that charges freely move along the field lines on the upper (and lower) boundaries. As a result, the gap becomes much thicker than MH04a,b; namely, $\Delta l_{\text{SG}} \sim 0.5 h_{\text{m}} \varpi_{\text{LC}}$, which gives $\epsilon < 10^{-3}$. Therefore, we can neglect the cross-field motion and justify the constancy of ρ/B in the outer region of the gap, where pair creation is negligible. In the inner magnetosphere, ρ_{eff}/B becomes approximately a negative constant, owing to the discharge of the copiously created pairs. Because of this negativity of ρ_{eff}/B, a positive $E_\parallel$ is exerted. For a super-GJ solution, we obtain $j_{\text{e}} + j_{\text{ion}} \sim 0.9 > \rho_{\text{eff}}/(\Omega B/2\pi)$, which guarantees the equipotentiality of the boundaries. For a sub-GJ solution, a problem may occur regarding the equipotentiality; nevertheless, we are not interested in this kind of solutions.

It is noteworthy that the electric current induced by a negative $E_\parallel$ contradicts with the global current patterns if the gap is located near the last-open field line. (No return current sheets on the last-open field line is supposed in slot gap models.) This situation is illustrated in figure 11 (see also the right panel of fig. 1). The current continuity suggests that a positive $E_\parallel$ should be exerted in the particle accelerator if it is located near the last-open field line.

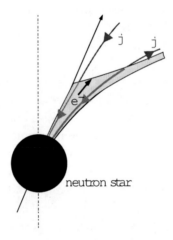

Figure 11. Current (downward arrows) derived in the inner-slot-gap (shaded region), and the global current patterns (arrows in non-shaded region) induced by the EMF on the spinning neutron-star surface. From Hirotani (2000b).

In short, whether the gap solution becomes MH04a way (with a negative $E_\parallel$ as an outward extension of the polar-cap model) or this-work way (with a positive $E_\parallel$ as an inward extension of the outer-gap model) entirely depends on the transfield thickness and on the resultant ρ_{eff}/B variation along the field lines. If $\Delta l_{\text{SG}} \sim r_*/10$ holds in the outer magnetosphere, ρ_{eff}/B could be a small positive constant by the cross-field motion of charges (without pair creation); in this case, the current is slightly sub-GJ with electron emission from the neutron star surface, as MH04a,b suggested. On the other hand, if $\Delta l_{\text{SG}} > \varpi_{\text{LC}}/40$ holds in the outer magnetosphere, ρ_{eff}/B takes a small negative value by the discharge of the created pairs (see fig. 10); in this case, the current is super-GJ with ion emission from the surface, as demonstrated in this chapter. Since no studies have ever successfully constrained the gap transfield thickness, there is room for further investigation on this issue.

The author wishes to express his gratitude to Drs. J. G. Kirk, A. K. Harding, B. Rudak, S. Shibata, and K. S. Cheng for fruitful discussion. This work is supported by the Theoretical Institute for Advanced Research in Astrophysics (TIARA) operated under Academia Sinica and the National Science Council Excellence Projects program in Taiwan administered through grant number NSC 94-2752-M-007-001. Also, this work is partly supported by KBN through the grant 2P03D.004.24 to B. Rudak, which enabled the author to use the MEDUSA cluster at CAMK Toruń.

References

[1] Becker, W., Trümper, J. 1997, *A&A* **326**, 682

[2] Beskin, V. S., Istomin, Ya. N., Par'ev, V. I. 1992, *Sov. Astron.* **36**(6), 642 (BIP92)

[3] Cheng, K. S., Ho, C., Ruderman, M., 1986a, *ApJ*, **300**, 500 (CHR86a)

[4] Cheng, K. S., Ho, C., Ruderman, M., 1986b, *ApJ*, **300**, 522 (CHR86b)

[5] Cheng, K. S., Zhang, L. 1996, *ApJ* **463**, 271

[6] Cheng, K. S., Ruderman, M., Zhang, L. 2000, *ApJ*, **537**, 964 (CRZ00)

[7] Chiang, J., Romani, R. W. 1992, *ApJ*, **400**, 629

[8] Chiang, J., Romani, R. W. 1994, *ApJ*, **436**, 754

[9] Comella, J. M., Craft, H. D., Lovelace, R. V. E., Sutton, J. M., Tayler, G. L. 1969, *Nature* **221**, 453.

[10] Daugherty, J. K., Harding, A. K. 1982, *ApJ*, **252**, 337 (DH82)

[11] Daugherty, J. K., Harding, A. K. 1996a, *ApJ*, **458**, 278 (DH96a)

[12] Daugherty, J. K., Harding, A. K. 1996b, *A& AS*, **120**, 107 (DH96a)

[13] Dermer, C. D., Sturner, S. J. 1994, *ApJ*, **420**, L75

[14] Fawley, W. M., Arons, J., Scharlemann, E. T. 1977, *ApJ* **217**, 227 (FAS77)

[15] Fierro, J. M., Michelson, P. F., Nolan, P. L., Thompson, D. J., 1998, *ApJ* **494**, 734

[16] Goldreich, P. Julian, W. H. 1969, *ApJ*. **157**, 869

[17] Harding, A. K., Tademaru, E., Esposito, L. S. 1978, *ApJ*, **225**, 226

[18] Higgins, M. G., Henriksen R. N. 1997, *MNRAS* **292**, 934

[19] Higgins, M. G., Henriksen R. N. 1998, *MNRAS* **295**, 188

[20] Hirotani, K. 2000a, *MNRAS* **317**, 225 (Paper IV)

[21] Hirotani, K. 2000b, *PASJ* **52**, 645 (Paper VI)

[22] Hirotani, K. 2001, *ApJ* **549**, 495 (Paper V)

[23] Hirotani, K. Okamoto, I., 1998, *ApJ*, **497**, 563 (Paper 0)

[24] Hirotani, K. Shibata, S., 1999a, *MNRAS* **308**, 54 (Paper I)

[25] Hirotani, K. Shibata, S., 1999b, *MNRAS* **308**, 67 (Paper II)

[26] Hirotani, K. Shibata, S., 1999c, *PASJ* **51**, 683 (Paper III)

[27] Hirotani, K. Shibata, S., 2001a, *MNRAS* **325**, 1228 (Paper VII)

[28] Hirotani, K. Shibata, S., 2001b, *ApJ* **558**, 216 (Paper VIII)

[29] Hirotani, K. Shibata, S., 2002, *ApJ* **564**, 369 (Paper IX)

[30] Hirotani, K., Harding, A. K., Shibata, S., 2003, *ApJ* **591**, 334 (Paper X)

[31] Hirotani, K. 2006a, submitted to *ApJ* (Paper XI)

[32] Hirotani, K. 2006b, Mod. Phys. Lett. A. (Brief Review) in press

[33] Jackson, J. D. 1962, *Classical electrodynamics* (New York: John Wiley & Sons), 591

[34] Jones, P. B. 1985, *Phys. Rev. Lett.*, **55**, 1338

[35] A. D. Kaminker, D. G. Yakovlev, O. Y. Gnedin 2002, *A & A* **383**, 1076

[36] Kanbach, G., Arzoumanian, Z., Bertsch, D. L., Brazier, K. T. S., Chiang, J., Fichtel, C. E., Fierro, J. M., Hartman, R. C., et al. 1994, *A & A* **289**, 855

[37] Kaspi, V. M., Lackey, J. R., Manchester, R. N., Bailes, M., Pace, R. 2000, *ApJ* **528**, 445

[38] Kuiper, L., Hermsen, W., Krijger, J. M., Bennett, K., Carramiana, A., Schönfelder, V., Bailes, M., Manchester, R. N. 1999, *A& A* **351**, 119

[39] Kuiper, L., Hermsen, W., Verbunt, F., Thompson, D. J., Stairs, I. H., Lyne, A. G., Strickman M. S., Cusumano, G. 2000, *A& A* **359**, 615

[40] Lense, J., Thirring, H. 1918, Phys. Z.19, 156. Translated by Mashhoon, B., Hehl, F.W., Theiss, D.S. 1984, *Gen. Relativ. Gravit.* **16**, 711

[41] Mayer-Hasselwander, H. A., Bertsch, D. L., Brazier, T. S., Chiang, J., Fichtel, C. E., Fierro, J. M., Hartman, R. C., Hunter, S. D. 1994, *ApJ* **421**, 276

[42] Mestel, L. 1971, *Nature Phys. Sci.*, **233**, 149

[43] Mestel, L., Robertson, J. A., Wang, Y. M., Westfold, K. C., 1985, *MNRAS* **217**, 443

[44] Muslimov, A. G., Tsygan, A. I. 1992, *MNRAS*, **255**, 61

[45] Muslimov, A. G., Harding, A. K. 2003 *ApJ* **588**, 430

[46] Muslimov, A. G., Harding, A. K. 2004a *ApJ* **606**, 1143

[47] Muslimov, A. G., Harding, A. K. 2004b *ApJ* **617**, 471

[48] Neuhauser, D., Langanke, K., Koonin, S. E., 1986, *Phys. Rev.* **A33**, 2084

[49] Neuhauser, D., Koonin, S. E., Langanke, K., 1986, *Phys. Rev.* **A36**, 4163

[50] Nolan, P. L., Arzoumanian, Z., Bertsch, D. L., Chiang, J., Fichtel, C. E., Fierro, J. M., Hartman, R. C., Hunter, S. D., et al. 1993, *ApJ* **409**, 697

[51] Pavlov, G. G., Zavlin, V. E., Sanwal, D., Burwitz, V., Garmire, G. P. 2001, *ApJ*, **552**, L129

[52] Ramanamurthy, P. V., Bertsch, D. L., Dingus, L., Esposito, J. A., Fierro, J. M., Fichtel, C. E., Hunter, S. D., Kanbach, G., et al. 1995, *ApJ* **447**, L109

[53] Ramanamurthy, P. V., Fichtel, C. E., Kniffen, D. A., Sreekumar, P., Thompson, D. J. 1996, *ApJ* **458**, 755

[54] Romani, R. W. 1996, *ApJ*, **470**, 469 (R96)

[55] Romani, R. W., Yadigaroglu, I. A. 1995, *ApJ* **438**, 314 (RY95)

[56] Scharlemann, E. T., Arons, J., Fawley, W. T., 1978, *ApJ*, **222**, 297 (SAF78)

[57] Shibata, S. 1995, *MNRAS* **276**, 537

[58] Shibata, S. 1997, *MNRAS* **287**, 262

[59] Staelin, D. H., Reifenstein E. C. 1968, *Science* **162**, 148.

[60] Sturner, S. J., Dermer, C. D., Michel, F. C. 1995, *ApJ* **445**, 736

[61] Takahashi, M., Nitta, S., Tatematsu, Y., Tomimatsu, A. 1990, *ApJ* **363**, 206

[62] J. Takata, S. Shibata, K. Hirotani, 2004, *MNRAS* **354**, 1120 (TSH04)

[63] J. Takata, S. Shibata, K. Hirotani, H. K. Chang 2006, *MNRAS* **366**, 1310 (TSHC06)

[64] Tennant, A. F., Becker, W., Juda, M., Elsner, R. F., Kolodziejczak, J. J. , Murray, S. S., O'Dell, S. L., Paerels, F., Swartz, D. A., Shibazaki, N.,Weisskopf, M. C. 2001, *ApJ* **554**, L173

[65] Thompson, D. J., Bailes, M., Bertsch, D. L., Esposito, J. A., Fichtel, C. E., Harding, A. K., Hartman, R. C., Hunter, S. D. 1996, *ApJ* **465**, 385

[66] Thompson, D. J., Bailes, M., Bertsch, D. L., Cordes, J., D'Amico, N., Esposito, J. A., Finley, J., Hartman, R. C., et al. 1999, *ApJ* **516**, 297

[67] Yabe, T., Aoki, T. 1991 *Comput. Phys. Commun.*, **66**, 219

[68] Yabe, T, Xiao, F., Utsumi, T. 2001, *J. Comput. Phys.* **169**, 556

[69] Zhang, L. Cheng, K. S. 1997, *ApJ* **487**, 370 (ZC97)

[70] R. L. Znajek, 1977, *MNRAS* **179**, 457

In: Pulsars: Discoveries, Functions and Formation
Editor: Peter A. Travelle, pp. 161-186

ISBN: 978-1-61122-982-0
© 2011 Nova Science Publishers, Inc.

Chapter 8

ACCRETION-DRIVEN MILLISECOND X-RAY PULSARS

Rudy Wijnands[*]
Astronomical Institute "Anton Pannekoek",
Kruislaan 403, NL-1098 SJ Amsterdam

Abstract

I present an overview of our current observational knowledge of the six known accretion-driven millisecond X-ray pulsars. A prominent place in this review is given to SAX J1808.4–3658; it was the first such system discovered and currently four outbursts have been observed from this source, three of which have been studied in detail using the *Rossi X-ray Timing Explorer* satellite. This makes SAX J1808.4–3658 the best studied example of an accretion-driven millisecond pulsar. Its most recent outburst in October 2002 is of particular interest because of the discovery of two simultaneous kilohertz quasi-periodic oscillations and nearly coherent oscillations during type-I X-ray bursts. This is the first (and so far only) time that such phenomena are observed in a system for which the neutron star spin frequency is exactly known. The other five systems were discovered within the last three years (with IGR J00291+5934 only discovered in December 2004) and only limited results have been published.

1. Introduction

Ordinary pulsars are born as highly-magnetized (B $\sim 10^{12}$ G), rapidly rotating (P ~ 10 ms) neutron stars which spin down on timescales of 10 to 100 million years due to magnetic dipole radiation. However, a number of millisecond (P < 10 ms) radio pulsars is known with ages of billions of years and weak (B $\sim 10^{8-9}$ G) surface magnetic fields. Since many of these millisecond pulsars are in binaries, it has long been suspected (see, e.g., Bhattacharya & van den Heuvel 1991 for an extended review) that the neutron stars were spun up by mass transfer from a stellar companion in a low-mass X-ray binary (LMXB), but years of searching for coherent millisecond pulsations in LMXBs failed to yield a detection (Vaughan et al. 1994 and references therein). The launch of the NASA *Rossi X-ray Timing*

---

[*]E-mail address: R.A.D.Wijnands@uva.nl; Tel: +31 20 525 7206; Fax: +31 20 525 7484

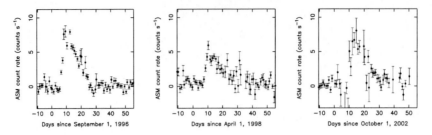

Figure 1. The *RXTE*/ASM light curves of SAX J1808.4–3658 during the September 1996 outburst (left), the April 1998 outburst (middle) and the October 2002 outburst (right). These light curves were made using the public ASM data available at http://xte.mit.edu/ASM_lc.html. The count rates are for the 2–12 keV energy range and are daily averages.

Explorer (*RXTE*) brought the discovery of kilohertz quasi-periodic oscillations (kHz QPOs; Strohmayer et al. 1996; Van der Klis et al. 1996) as well as nearly coherent oscillations ('burst oscillations') during type-I X-ray bursts in a number of LMXBs (e.g., Strohmayer et al. 1996), providing tantalizingly suggestive evidence for weakly magnetic neutron stars with millisecond spin periods (see Van der Klis 2000, 2004 and Strohmayer & Bildsten 2003 for more details about kHz QPOs and burst oscillations in LMXBs).

In April 1998 the first accretion-driven millisecond X-ray pulsar (SAX J1808.4–3658) was discovered (Wijnands & van der Klis 1998a) proving that indeed neutron stars in LMXBs can spin very rapidly. This conclusion was further strengthened by the discovery of four additional systems in 2002 and 2003 (Markwardt et al. 2002a, 2003a, 2003b, Galloway et al. 2002), and recently, in December 2004, with the discovery of IGR J00291+5934 as a millisecond X-ray pulsar (Markwardt et al. 2004a). Here, I will give a brief summary of our current observational knowledge of those accretion-driven millisecond X-ray pulsars. Preliminary versions of this review were published by Wijnands (2004a, 2004b).

2.   SAX J1808.4–3658

2.1.   The September 1996 Outburst

In September 1996, a new X-ray transient and LMXB was discovered with the Wide Field Cameras (WFCs) aboard the Dutch-Italian *BeppoSAX* satellite and the source was designated SAX J1808.4–3658 (In 't Zand et al., 1998). Three type-I X-ray bursts were detected, demonstrating that the compact object in this system is a neutron star. From those bursts, a distance estimate of 2.5 kpc was determined (In 't Zand et al. 1998, 2001). The maximum luminosity during this outburst was $\sim 10^{36}$ erg s^{-1}, significantly lower than the peak outburst luminosity of 'classical' neutron star transients (which typically can reach a luminosity of 10^{37} to 10^{38} ergs s^{-1}). This low peak luminosity showed that the source was part of the growing group of faint neutron-star X-ray transients (Heise et al., 1999). The outburst continued for about three weeks (see Fig. 1) after which the source was thought to have returned to quiescence. However, it was found (Revnivtsev 2003) that the source

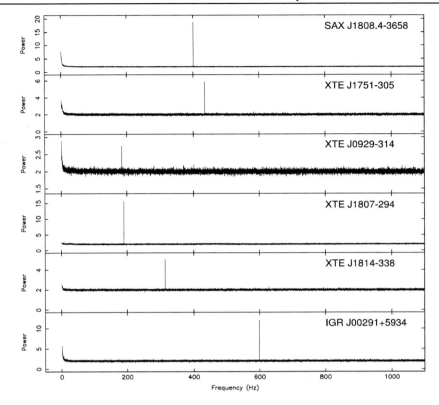

Figure 2. Examples of power spectra for each of the six currently known millisecond X-ray pulsars showing the pulsar spikes.

was detected on October 29, 1996 (using slew data obtained with the proportional counter array [PCA] aboard *RXTE*) with a luminosity of about a tenth of the outburst peak luminosity. This demonstrates that six weeks after the main outburst the source was still active (possibly only sporadically), which might indicate that at the end of this outburst the source behaved in a manner very similar to what was seen during its 2000 and 2002 outbursts (see § 2.3. and § 2.4.).

After it was found that SAX J1808.4–3658 harbors a millisecond pulsar (§ 2.2.), the three observed X-ray bursts seen with *BeppoSAX*/WFC were scrutinized for potential burst oscillations (In 't Zand et al., 2001). A marginal detection of a 401 Hz oscillation was made in the third burst. This result suggested that the burst oscillations observed in the other, non-pulsating, neutron-star LMXBs occur indeed at their neutron-star spin frequencies. This result has been confirmed by the recent detection of burst oscillations during the 2002 outburst of SAX J1808.4–3658 (§ 2.4.2.).

2.2.   The April 1998 Outburst

On April 9, 1998, *RXTE*/PCA slew observations indicated that SAX J1808.4–3658 was active again (Marshall 1998; see Fig. 1 for the *RXTE*/ASM light curve during this outburst). Using public TOO observations of this source from April 11, it was discovered (Wijnands

& Van der Klis 1998a) that coherent 401 Hz pulsations (Fig. 2) were present in the persistent X-ray flux of the source, making it the first accretion-driven millisecond X-ray pulsar discovered. After this discovery, several more public *RXTE* observations were made (using the PCA) which were used by several groups to study different aspects of the source. I will briefly mention those results and I point to references for the details.

A detailed analysis of the coherent timing behavior (Chakrabarty & Morgan 1998) showed that the neutron star was in a tight binary with a very low-mass companion star in a $\sim$2-hr orbital period. Due to the limited amount of data obtained during this outburst, only an upper limit of $< 7 \times 10^{-13}$ Hz s^{-1} could be obtained on the pulse-frequency derivative (Chakrabarty & Morgan 1998). Studies of the X-ray spectrum (Gilfanov et al., 1998; Heindl & Smith 1998; see also Gierlinski et al., 2002 and Poutanen & Gierlinski 2003) and the aperiodic rapid X-ray variability (Wijnands & van der Klis 1998b; see also Van Straaten et al., 2005) showed an object that, apart from its pulsations, is remarkably similar to other LMXBs with comparable luminosities (the atoll sources). There is apparent modulation of the X-ray intensity at the orbital period, with a broad minimum when the pulsar is behind the companion (Chakrabarty & Morgan 1998; Heindl & Smith 1998). Cui et al., (1998) and Ford (2000) reported on the harmonic content, energy dependency, and soft phase lag of the pulsations. The main result of those studies is that the low-energy pulsations lag the high-energy ones by as much as $\sim$200 μs ($\sim$8% of the pulsation period; see Cui et al., [1998], Ford [2000], and Poutanen & Gierlinski [2003] for possible explanations for these soft lags).

Another interesting aspect is that the source first showed a steady decline in X-ray flux, which after 2 weeks suddenly accelerated (Gilfanov et al., 1998; Cui et al., 1998; Fig. 8). This behavior has been attributed to the fact that the source might have entered the 'propeller regime' in which the accretion is centrifugally inhibited (Gilfanov et al. 1998). However, after the onset of the steep decline the pulsations could still be detected (Cui et al., 1998) making this interpretation doubtful. A week after the onset of this steep decline, the X-ray flux leveled off (Cui et al., 1998; Wang et al., 2001), but as no further *RXTE*/PCA observations were made, the X-ray behavior of the source at the end of the outburst remained unclear. The source might have displayed a similar long-term episode of low-luminosity activity as seen at the end of its 2000 and 2002 outbursts (see § 2.3. and § 2.4.).

SAX J1808.4–3658 was not only detected and studied in X-rays but also in the optical, IR, and in radio bands. The optical/IR counterpart of SAX J1808.4–3658 (later named V4580 Sgr; Kazarovets et al., 2000) was first discovered by Roche et al., (1998) and subsequently confirmed by Giles et al. (1998). A detailed study of the optical behavior during this outburst was reported by Giles et al. (1999) and Wang et al. (2001). Both papers reported that the peak V magnitude of the source was $\sim$16.7 and the source decayed in brightness as the outburst progressed. The brightness of the source leveled off at around V $\sim$ 18.5 (I $\sim$ 17.9) about $\sim$2 weeks after the peak of the outburst. It stayed at this level for at least several weeks before it further decreased in brightness. This behavior suggests that the source was indeed still active for a long period after the main outburst.

It was also reported (Giles et al., 1999) that the optical flux was modulated at the 2-hr orbital period of the system. Modeling the X-ray and optical emission from the system using an X-ray-heated accretion disk model yielded a Av of 0.68 and an inclination of cos i = 0.65 (Wang et al., 2001), resulting in a mass of the companion star of 0.05–0.10 solar masses. During some of the IR observations, the source was too bright to be consistent with emission from the disk or the companion star, even when considering X-ray heating. This IR excess might be due to synchrotron processes, likely related to an outflow or ejection of matter (Wang et al., 2001). Such an ejection event was also confirmed by the discovery of the radio counterpart (Gaensler et al., 1999). The source was detected with a 4.8 GHz flux of ~0.8 mJy on 1998 April 27, but it was not detected at earlier or later epochs.

2.3.   The January 2000 Outburst

On January 21, 2000, SAX J1808.4–3658 was again detected (Wijnands et al. 2001) with the *RXTE*/PCA at a flux level of ~10–15 mCrab (2–10 keV), i.e. about a tenth of the peak fluxes observed during the two previous outbursts. Using follow-up *RXTE*/PCA observations, it was found that the source exhibited low-level activity for several months (Wijnands et al., 2001). Due to solar constraints the source could not be observed before January 21 but likely a true outburst occurred before that date and we only observed the end stages of this outburst. This is supported by the very similar behavior of the source observed near the end of its 2002 October outburst (see § 2.4.; Fig. 5).

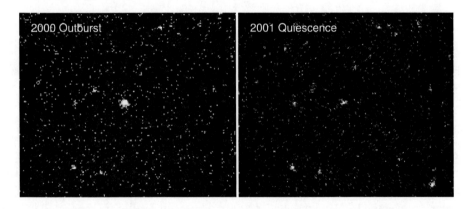

Figure 3. The *XMM-Newton* images of the field containing SAX J1808.4–3658 during its 2000 outburst (left panel; Wijnands 2003) and when the source was in quiescence (in 2001; right panel; see Campana et al. 2002). Clearly, SAX J1808.4–3658 (the source in the middle of the image) was brighter (albeit if only by a factor of a few) during the 2000 outburst observation than during the quiescent observation.

During the 2000 outburst, SAX J1808.4–3658 was observed (using *RXTE*/PCA) on some occasions at luminosities of ~10^{35} ergs s^{-1}, but on other occasions (a few days earlier or later) it had luminosities of ~10^{32} ergs s^{-1} (as seen during *BeppoSAX* and *XMM-Newton* observations; Wijnands et al., 2002, Wijnands 2003; see Fig. 3 left panel). This demonstrates that the source exhibited extreme luminosity swings (a factor of >1000) on timescales of days. During the *RXTE* observations, it was also found that on several occasions the source exhibited strong (up to 100 % r.m.s. amplitude) violent flaring behavior with a repetition frequency of about 1 Hz (Van der Klis et al., 2000; Fig. 9). During this episode of low-level activity, the pulsations at 401 Hz were also detected.

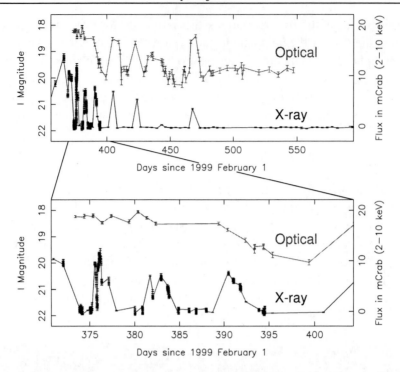

Figure 4. The *RXTE*/PCA (Wijnands et al., 2001) and the optical (I band) light curves (Wachter et al., 2000) of SAX J1808.4–3658 as observed during its 2000 outburst. The optical data were kindly provided by Stefanie Wachter.

The source was again detected in optical, albeit at a lower brightness than during the 1998 outburst (Wachter & Hoard 2000). This is consistent with the lower X-ray activity seen for the source. The source was frequently observed during this outburst and preliminary results were presented by Wachter et al., (2000). The main results are presented in Figure 4 (reproduced with permission from Stefanie Wachter). The optical and X-ray brightness of the source are correlated at the end of the outburst, although one optical flare (around day 435–440 in Fig. 4) was not accompanied by an X-ray flare. However, the optical and X-ray observations were not simultaneous, which means that a brief (around a few days) X-ray flare could have been missed. During the earlier stages of the outburst, the X-ray and the optical behavior of the source were not correlated (Fig. 4 lower panel): the source is highly variable in X-rays, but quite stable in optical with only low amplitude variations. This stable period in the optical is very similar to the episode of stable optical emission in the late stages of the 1998 outburst, suggesting this is typical behavior for this source.

2.4.   The October 2002 Outburst

In 2002 October, the fourth outburst of SAX J1808.4–3658 was detected (Markwardt et al., 2002b), immediately launching an extensive *RXTE*/PCA observing campaign. The main results are summarized below.

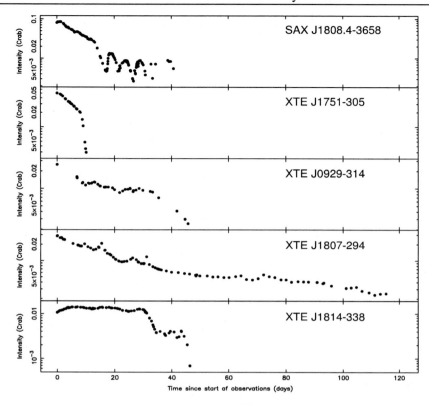

Figure 5. The *RXTE*/PCA light curves of five of the six accretion-driven millisecond X-ray pulsars. The data for SAX J1808.4-3658 was obtained during its 2002 outburst. The data were taken from van Straaten et al. (2005), except for XTE J1807–294 which were taken from Linares et al. (2005).

2.4.1.   The X-ray Light Curve

The *RXTE*/PCA light curve for this outburst is shown in Figure 5 (see Fig. 1 for the ASM light curve). During the first few weeks, the source decayed steadily, until the rate of decline suddenly increased, in a manner similar to what was observed during the 1998 outburst (see § 2.2.). During both the 1998 and 2002 outbursts, the moment of acceleration of the decline occurred at about two weeks after the peak of the outburst. Approximately five days later the X-ray count rate rapidly increased again until it reached a peak of about a tenth of the outburst maximum. After that the source entered a state in which the count rate rapidly fluctuated on time scales of days to hours, very similar to the 2000 low-level activity (see § 2.3.). The 2002 outburst light curve is the most detailed one seen for this source and it exhibits all features seen during the previous three outbursts of the source (the initial decline, the increase in the decline rate, the long-term low-level activity), demonstrating that this behavior is typical for this source.

2.4.2. The X-ray Bursts and the Burst Oscillations

During the first five days of the outburst, four type-I X-ray bursts were detected. Burst oscillations were observed during the rise and decay of each burst, but not during the peak (Chakrabarty et al. 2003). The frequency in the burst tails was constant and identical to the spin frequency, while the oscillation in the burst rise showed evidence for a very rapid frequency drift of up to 5 Hz. This frequency behavior and the absence of oscillations at the peak of the bursts is similar to the burst oscillations seen in other, non-pulsating neutron star LMXBs, demonstrating that indeed the burst-oscillations occur at the neutron-star spin frequency in all sources. As a consequence, the spin frequency is now known for 18 LMXBs (12 burst-oscillations sources and 6 pulsars) with the highest spin frequency being 619 Hz. The sample of burst-oscillation sources was used to demonstrate that neutron stars in LMXBs spin well below the break-up frequency for neutron stars. This could suggest that the neutron stars are limited in their spin frequencies, possible due to the emission of gravitational radiation (Chakrabarty et al., 2003; Chakrabarty 2004).

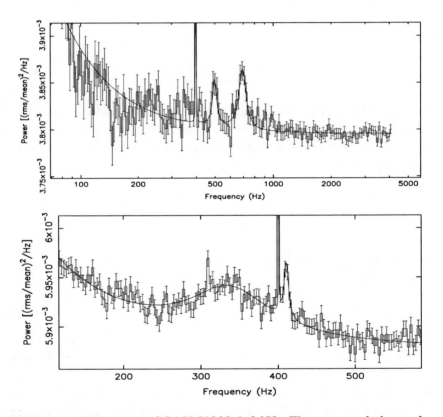

Figure 6. The power spectrum of SAX J1808.4–3658. The top panel shows the two simultaneous kHz QPOs discovered during its 2002 outburst. The bottom panel shows the enigmatic 410 Hz QPO also seen during this outburst. The figures are adapted from Wijnands et al., (2003).

2.4.3.   The kHz QPOs

Wijnands et al. (2003) reported on the discovery of two simultaneous kHz QPOs during the peak of the outburst with frequencies of $\sim$700 and $\sim$500 Hz (Fig. 6 top panel). This was the first detection of twin kHz QPOs in a source with a known spin-frequency. The frequency separation of those two kHz QPOs is only $\sim$200 Hz, significantly below the 401 Hz expected in the beat-frequency models proposed to explain the kHz QPOs. Therefore, those models are falsified by the discovery of kHz QPOs in SAX J1808.4–3658. The fact that the peak separation is approximately half the spin frequency suggests that the kHz QPOs are indeed connected to the neutron-star spin frequency, albeit in a way not predicted by any existing model at the time of the discovery. The lower-frequency kHz QPO was only seen during the peak of the outburst (October 16, 2002) but the higher-frequency kHz QPO could be traced throughout the main part of the outburst (Wijnands et al., 2003). In addition to the twin kHz QPOs, a third kHz QPO was found with frequencies ($\sim$410 Hz) just exceeding the pulse frequency (Fig. 6 bottom panel; Wijnands et al., 2003). The nature of this QPO is unclear but it might be related to the side-band kHz QPO seen in several other sources (Jonker et al., 2000).

Wijnands et al., (2003) pointed out that there appear to exist two classes of neutron-star LMXBs: the 'fast' and the 'slow' rotators. The fast rotators have spin frequencies $>\sim$400 Hz and the frequency separation between the kHz QPOs is roughly equal to half the spin frequency. In contrast, the slow rotators have spin frequencies below $<\sim$400 Hz and a frequency separation roughly equal to the spin frequency. These latest kHz QPO results have spurred new theoretical investigations into the nature of kHz QPO, involving spin induced resonance in the disk (e.g., Wijnands et al., 2003; Kluzniak et al. 2004; Lee et al., 2004; Lamb & Miller 2004; Kato 2004).

2.4.4.   The Low-Frequency QPOs

During the peak of the outburst and in its subsequent decay, broad-noise and QPOs with frequencies between 10 and 80 Hz were detected in the power spectra (Fig. 7). Similar phenomena have been observed in other non-pulsating systems and are likely to be related to the noise components seen in SAX J1808.4–3658. Van Straaten et al. (2004, 2005) have studied the broad-band power spectra (including the noise components, the low-frequency QPOs, and the kHz QPOs) of SAX J1808.4–3658 in detail as well as the frequency correlations between the different power-spectral components. Interestingly, using those frequency correlations, van Straaten et al. (2004, 2005) suggested that the higher-frequency kHz QPO could also be identified during the 1998 outburst but at the lowest frequencies found so far in any kHz QPO source (down to $\sim$150 Hz). Previous work (Wijnands & van der Klis 1998b) on the aperiodic timing features of SAX J1808.4–3658 during its 1998 outburst had already found these features but they could not be identified as the higher-frequency kHz QPO due to their low frequency and broad character.

Van Straaten et al. (2004, 2005) also compared the results of SAX J1808.4–3658 with those obtained for other non-pulsating neutron-star LMXBs. In those other sources, the frequencies of the variability components follow an universal scheme of correlations. The correlations observed for SAX J1808.4–3658 are similar but they show a shift in the frequencies of the kHz QPOs. It is unclear what physical mechanism(s) underlies this differ-

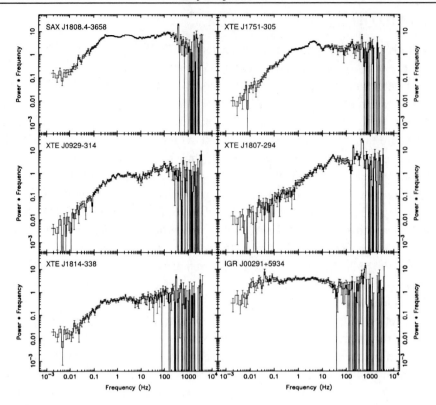

Figure 7. Examples of the aperiodic timing features seen in the six millisecond pulsars. For SAX J1808.4–3658 we show a power spectrum obtained during its 1998 outburst.

ence among sources (van Straaten et al., 2004, 2005).

During the 1998 and 2002 outbursts of SAX J1808.4–3658, the source exhibited similar X-ray fluxes. However, at similar flux levels, the characteristic frequencies observed during the 1998 outburst are much lower (by a factor of~10) than during the 2002 outburst (van Straaten et al., 2005; see Fig. 8). Again it is unclear what causes this huge difference between the two outbursts but it might be related to the 'parallel track' phenomena observed for the kHz QPOs in the non-pulsating neutron-star LMXBs (e.g., van der Klis 2000).

2.4.5.   The Violent 1 Hz Flaring

Violent flaring was observed on many occasions at a ~1 Hz repetition frequency during the late stages of the 2002 outburst (Fig. 9), similar to what had been observed during the 2000 outburst. This proves that also this violent flaring is a recurrent phenomenon and can likely be observed every time the source is in this prolonged low-level activity state. Preliminary results presented in Figure 9 (right panels) show examples of power spectra obtained during the end stages of the 2002 outburst. During certain observations the 1 Hz QPO is rather narrow and its first overtone can be seen clearly (Fig. 9 top right panel). During other observations, the 1 Hz QPO is much broader and its wings blend with the first overtone (Fig. 9 middle two right panels). In addition to the 1 Hz QPO, other QPOs

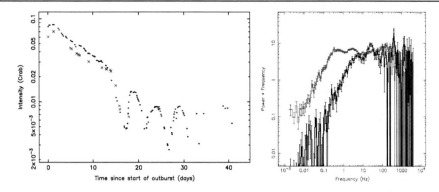

Figure 8. *Left*: The *RXTE*/PCA light curves of SAX J1808.4–3658 during its 1998 (crosses) and 2002 (dots) outbursts. The data were taken from van Straaten et al. (2005). *Right*: A comparison of the power spectrum for SAX J1808.4–3658 obtained at roughly the same flux levels (∼0.044 Crab) during the 1998 outburst (light gray; obtained on April 16, 1998) with that obtained during the 2002 outburst (black; October 23, 2002).

around 30–40 Hz are sometimes seen (see also van Straaten et al., 2005). It is unclear if this 30–40 Hz QPO is related to the low-frequency QPOs discussed in § 2.4.4. or if it is due to a different mechanism. During certain observations the 1 Hz QPO becomes very broad, turning into a band-limited noise component (Fig. 9 bottom right panel). The mechanism behind these violent flares is not yet known and a detailed analysis of this phenomenon is in progress.

2.4.6. The Pulsations

The pulsations could be detected at all flux levels with an amplitude of 3% – 10%. There was no evidence for a 200.5 Hz subharmonic in the data (upper limit of 0.38% of the signal at 401 Hz; Wijnands et al. 2003) confirming the interpretation of 401 Hz as the pulsar spin frequency. A detailed analysis and discussion of the coherent timing analysis will be presented by Morgan et al. (2006, in preparation).

2.4.7. Observations at Other Wavelengths

Rupen et al. (2002b) reported the detection of SAX J1808.4–3658 at radio wavelengths. On October 16, 2002, they found a 0.44-mJy source at 8.5 GHz and a day later, the source was detected at 0.3 mJy. Monard (2002) reported that on October 16, 2002, the optical counterpart was detectable again at magnitudes similar to those observed at the peak of the 1998 outburst.

2.5. SAX J1808.4–3658 in Quiescence

In quiescence, SAX J1808.4–3658 has been observed on several occasions with the *BeppoSAX* and *ASCA* satellites (Stella et al. 2000; Dontani et al., 2000; Wijnands et al., 2002). The source was very dim in quiescence, with a luminosity close to or lower than 10^{32} ergs

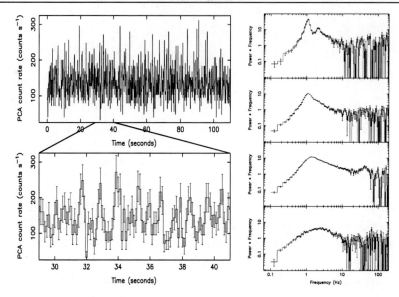

Figure 9. The violent 1 Hz flaring as observed during the 2000 and 2002 outbursts of SAX J1808.4-3658. The flaring can be seen in the light curve on the left panels and in the power spectra on the right panels.

s^{-1}. Due to the low number of source photons detected, these luminosities had large errors and no information could be obtained on the spectral shape or possible variability in quiescence. Due to the limited angular resolution of *BeppoSAX*, doubts were raised as to whether the source detected by this satellite was truly SAX J1808.4–3658 or an unrelated field source (Wijnands et al. 2002). Campana et al. (2002) reported on a quiescent observation of the source performed with *XMM-Newton* which resolved this issue. They detected the source at a luminosity of 5×10^{31} ergs s^{-1} and found that the field around SAX J1808.4–3658 is rather crowded with weak sources. Two such sources are relatively close to SAX J1808.4–3658 and might have conceivably caused a systematic positional offset during the *BeppoSAX* observations. Despite this fact it is very likely that SAX J1808.4–3658 was indeed detected during those *BeppoSAX* observations.

Using *XMM-Newton*, Campana et al. (2002) obtained enough photons to extract a quiescent X-ray spectrum, which was not dominated by the same thermal component seen in other quiescent neutron star transients; such a thermal component is thought to be due to the cooling of the neutron star in-between outbursts. However, the spectrum of SAX J1808.4–3658 was dominated by a power-law shaped component. The non-detection of the thermal component was used to argue that the neutron star was anomalously cool, possibly due to enhanced core cooling processes (Campana et al. 2002). It has been argued (Stella et al., 2000; Campana et al., 2002) that the propeller mechanism, which might explain (some of) the hard X-ray emission in quiescence, is not likely to be active since this mechanism is expected to stop operating at luminosities $<10^{33}$ ergs s^{-1}, because at those luminosities the source should turn on as a radio pulsar. Instead, it was proposed that the quiescent X-rays originate in the shock between the wind of a turned-on radio pulsar and the matter flowing out from the companion star (Stella et al., 2000; Campana et al., 2002). Di Salvo & Burderi

(2003) suggested that the quiescent X-rays could also be due to direct dipole radiation from the radio pulsar. Using simple accretion disk physics and the quiescent luminosity found by Campana et al. (2002), they determined that the magnetic field strength of the neutron star in SAX J1808.4–3658 should be in a quite narrow range of $(1 - 5) \times 10^8$ Gauss.

The quiescent optical counterpart of SAX J1808.4–3658 was studied by Homer et al. (2001). They reported that on August 10, 1999, the orbital modulation was still present in white light observations (estimated V magnitude of $\sim$20), with a semi-amplitude of $\sim$6%. It has the same phasing and approximately sinusoidal modulation as seen during outburst, and with photometric minimum when the pulsar is behind the companion star. During observations taken in July 2000 the quiescent counterpart was even fainter and no significant orbital modulation could be detected. Using these results, it has been suggested that the optical properties of SAX J1808.4–3658 in quiescence are evidence of an active radio pulsar (Burderi et al. 2003). Campana et al. (2004) reported on the first optical spectrum of this source during its quiescent state. They concluded that a very high irradiating luminosity, a factor of $\sim$100 larger than directly observed from the X-rays, must be present in the systems, which was suggested to be derived from a rotation-powered neutron star. If true, a pulsating radio source might be expected, but a search at 1.4 GHz could not detected the source (Burgay et al. 2003). This could be due to the effects of free-free absorption and searches at higher frequencies to limit these effects might still yield a pulsating radio source during the quiescent state of SAX J1808.4-3658.

3. XTE J1751–305

3.1. The 2002 Outburst

The second accretion-driven millisecond pulsar (XTE J1751–305) was discovered on April 3, 2002 (Markwardt et al., 2002a). Its spin frequency is 435 Hz (Fig. 2) and the neutron star is in a very small binary with an orbital period of only 42 minutes. The timing analysis of the pulsations gave a minimum mass for the companion star of 0.013 solar mass and a pulse-frequency derivative of $<3 \times 10^{-13}$ Hz s^{-1}. Assuming that the mass transfer in this binary system was driven by gravitational radiation, the distance toward the source could be constrained to at least 7 kpc and the orbital inclination to 30°–85°, resulting in a companion mass of 0.013–0.035 solar masses, suggesting a heated helium dwarf (Markwardt et al., 2002a). *Chandra* briefly observed the source, resulting in an arcsecond position (Markwardt et al. 2002a).

The source reached a peak luminosity of $>2 \times 10^{37}$ ergs s^{-1}, an order of magnitude brighter than the peak luminosity of SAX J1808.4–3658. However, the outburst was very short with an e-folding time of only $\sim$7 days (compared to $\sim$14 days for SAX J1808.4–3658; Fig. 5) resulting in a low outburst fluence of only $\sim$2.5 $\times 10^{-3}$ ergs cm^{-2} (Markwardt et al. 2002a). A potential re-flare was seen two weeks after the end of the outburst during which also a type-I X-ray burst was seen. Preliminary analysis of the burst indicated that the burst did not come from XTE J1751–305 but from another source in the field of view. This was later confirmed (In 't Zand et al., 2003) and the burst likely originated from the bright X-ray transient in Terzan 6. It was also determined that the transient in Terzan 6 could not have produced the re-flare (In 't Zand et al., 2003) suggesting that this re-flare could still have come from XTE J1751–305. However, van Straaten et al. (2005) suggested (based on a X-ray color study using *RXTE*/PCA observations) that this re-flare was emitted by one of the background sources and not by XTE J1751–305. Van Straaten et al. (2005) also investigated the aperiodic timing properties of the source (an example power spectrum is shown in Fig. 7) and the correlations between the characteristic frequencies of the observed power-spectral components. The frequency correlations were similar to those of the non-pulsating neutron-star LMXBs. In contrast with the results obtained for SAX J1808.4–3658 (§ 2.4.4.), no frequency shift was required for XTE J1751–305 to make the frequency correlations consistent with those of the non-pulsating sources. Using these correlations, van Straaten et al. (2005) suggested that the highest-frequency noise components in XTE J1751–305 are likely due to the same physical mechanisms as the kHz QPOs. They also investigated the correlations between the characteristic frequencies and the X-ray colors of the source and concluded that it did not behave like an atoll source.

A previous outburst in June 1998 was discovered using archival *RXTE*/ASM data (Markwardt et al., 2002a), suggesting a tentative recurrence time of ∼3.8 years. Miller et al. (2003) reported on high spectral resolution data of the source obtained with *XMM-Newton* to search for line features in the X-ray spectrum. However, they only detected a continuum spectrum dominated by a hard power-law shaped component (power-law index of ∼1.44) but with a 17% contribution to the 0.5–10 keV flux from a soft thermal (black-body) component with temperature of ∼1 keV. Gierlinski & Poutanen (2005) studied in detail the X-ray spectrum of the source during its 2002 outburst using the archival *RXTE* and *XMM-Newton* observations. They find that XTE J1751–305 exhibited very similar behavior as SAX J1808.4–3658 during its 1998 outbursts. They also find that the pulse profile cannot be described by a simple sinusoid, but that a second harmonic is needed (the peak-to-peak amplitude of the fundamental was found to be 4.5% but of the second harmonic only 0.15%). Gierlinski & Poutanen (2005) also report that a clear energy dependency of the pulse profile was observed and that the higher energy photons arrive earlier than the softer (this 'soft lag' reached ∼100 μs at about 10 keV, where it saturated). Searches for the optical and near-infrared counterparts were performed but no counterparts were found (Jonker et al. 2003), likely due to the high reddening toward the source. These non-detections did not constrain any models for the accretion disk or possible donor stars.

3.2. XTE J1751–305 in Quiescence

Recently, XTE J1751–305 was observed in quiescence using *Chandra* (Wijnands et al., 2005). Sadly, they could not detect the source in their ∼43 ksec observation, with 0.5–10 keV flux upper limits between 0.2 and 2.7×10^{-14} ergs s^{-1} cm^{-2} depending on assumed spectral shape, resulting in 0.5–10 keV luminosity upper limits of $0.2 - 2 \times 10^{32}$ $(d/\, 8\,\mathrm{kpc})^2$ ergs s^{-1}, with d the distance toward the source in kpc. Using simple accretion disk physics in combination with these luminosity upper limits, Wijnands et al. (2005) could constrain the magnetic field of the neutron star in this system to be less than $3 - 7 \times 10^8 \frac{d}{8\,\mathrm{kpc}}$ Gauss (depending on assumed spectral shape of the quiescent spectrum).

4. XTE J0929–314

4.1. The 2002 Outburst

The third accretion-driven millisecond X-ray pulsar XTE J0929–314 had already been detected with the *RXTE*/ASM on April 13, 2002 (Remillard 2002) but was only found to be harboring a millisecond pulsar with a pulsations frequency of 185 Hz (Fig. 2) on May 2nd when observations of the source were made using the *RXTE*/PCA (Remillard et al., 2002). Galloway et al. (2002) reported on the detection of the 44-min orbital period of the system which is remarkably similar to that of XTE J1751–305. A minimum mass of 0.008 solar mass was obtained for the companion star and a pulse-frequency derivative of $(-9.2 \pm 0.4) \times 10^{-14}$ Hz s^{-1}. Galloway et al. (2002) suggested that this spin down torque may arise from magnetic coupling to the accretion disk, a magneto-hydrodynamic wind, or gravitational radiation from the rapidly spinning neutron star. Assuming gravitational radiation as the driving force behind the mass transfer, Galloway et al. (2002) found a lower limit to the distance of 6 kpc. They also reported on the detection of a QPO at 1 Hz (Fig. 7). Full details of this QPO and the other aperiodic power-spectral components are presented by van Straaten et al. (2005). Just as they found for SAX J1808.4–3658, the frequency correlations for XTE J0929–314 were similar to those observed for the non-pulsating sources but with an offset in the frequencies of the highest-frequency components. These correlations allowed van Straaten et al. (2005) to identify those components as related to the kHz QPOs. Studying the correlated spectral and timing variability, they concluded that the behavior of XTE J0929–314 was consistent with that of an atoll source.

Juett et al. (2003) obtained high resolution spectral data using the *Chandra* gratings. Again the spectrum is well fitted by a power-law plus a black body component, with a power-law index of 1.55 and a temperature of 0.65 keV. Similar to XTE J1751–305, no emission or absorption features were found. No orbital modulation of the X-ray flux was found implying an upper limit on the inclination of 85°. Greenhill et al. (2002) reported the discovery of the optical counterpart of the system with a V magnitude of 18.8 on May 1st, 2002 (see also Cacella 2002). Castro-Tirado et al. (2002) obtained optical spectra of the source on May 6–8 in the range 350–800 nm and found emission lines from the C III - N III blend and H-alpha, which were superposed on a blue continuum. These optical properties are typical of X-ray transients during outburst. Rupen et al. (2002a) discovered the radio counterpart of the source using the VLA with 4.86 GHz flux of 0.3–0.4 mJy.

4.2. XTE J0929–314 in Quiescence

Recently, Wijnands et al. (2005) also observed XTE J0929–314 in its quiescent state with *Chandra*. For this source, they detected 22 source photons (in the energy range 0.3–8 keV) in ∼24.4 ksec of on-source time. This small number of photons detected did not allow for a detailed spectral analysis of the quiescent spectrum, but they could demonstrate that the spectrum is harder than a simple thermal emission (which might have been due to the cooling neutron star that has been heated during outbursts). Assuming a power-law spectral model for the time-averaged (averaged over the whole observation) X-ray spectrum, they obtained a power-law index of ∼1.8 and an unabsorbed X-ray flux of ∼6×10^{-15} ergs s^{-1} cm^{-2} (for the energy range 0.5–10 keV), resulting in a 0.5–10 keV X-ray luminosity of

$\sim 7 \times 10^{31}$ $(d/10 \text{ kpc})^2$ ergs s^{-1}, with d the distance in kpc. The thermal component usually seen in quiescent neutron star LMXBs could not be detected, with a maximum contribution to the 0.5–10 keV flux of $\sim 30\%$. Wijnands et al. (2005) also found that the quiescent count rate of XTE J0929–314 was variable at the 95% confidence level, but no conclusive evidence was found for associated spectral variability. The properties of XTE J0929–314 in its quiescent state are remarkably similar to that observed for SAX J1808.4–3658 (§ 2.5.) which might suggest that such behavior is common among accretion-driven millisecond X-ray pulsars. However, recent work on several other weak quiescent neutron-star X-ray binaries (e.g., Jonker et al., 2004a, b ; Tomsick et al., 2004), which do not exhibit pulsations during their X-ray outbursts, suggests that also such systems can resemble SAX J1808.4–3658 during their X-ray outbursts, suggests that also such systems can resemble SAX J1808.4–3658 during quiescent (i.e., they could be almost as faint and hard as SAX J1808.4–3658 in quiescence; see Wijnands et al., 2005 for a in-dept discussion). Similar to what they did for XTE J1751–305, Wijnands et al. (2005) could constrain the neutron-star magnetic field strength in XTE J0929–314 to be $< 3 \times 10^9 \frac{d}{10 \text{ kpc}}$ Gauss.

5. XTE J1807–294

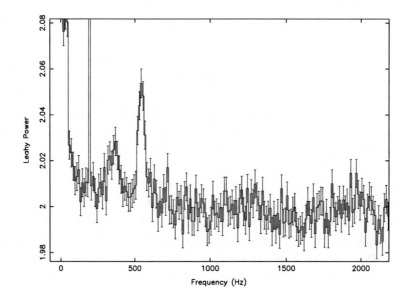

Figure 10. The power spectrum of XTE J1807–294 as obtained using *RXTE*/PCA. The two simultaneous kHz QPOs are clearly visible.

The fourth millisecond X-ray pulsar XTE J1807–294 with a frequency of 191 Hz, was discovered on February 21, 2003 (Markwardt et al., 2003a; Fig. 2). The peak flux was only 58 mCrab (2–10 keV, measured on February 21, 2003). The orbital period was determined (Markwardt et al. 2003c) to be ~ 40 minutes making it the shortest orbital period of all accretion-driven millisecond pulsars now known. Markwardt et al. (2003c) reported the best known position of the source based on a *Chandra* observation. Using the *RXTE*/PCA data, kHz QPOs have been detected for this system and it was found that the frequency separation between the two kHz QPOs was consistent with being equal to the neutron-star spin frequency (Linares , 2005; Fig. 10). This makes XTE J1807–294 consistent with the classification of Wijnands et al. (2003) of the neutron-star LMXBs into 'fast' and 'slow' rotators, with XTE J1807–294 a slow rotator. A detailed analysis of the correlations between the kHz QPOs and the low-frequency features (see Fig. 7) in this source is reported by Linares et al. (2005). The results of that analysis shows that also for XTE J1807–294 the frequency correlations are similar to those observed for the non-pulsating sources but with an offset in the frequencies of the highest-frequency components (similar to what was found for SAX J1808.4–3658 and possible XTE J0929–314; Van Straaten et al. 2005). Campana et al. (2003) reported on a *XMM-Newton* observation of this source taken on March 22, 2003. Assuming a distance of 8 kpc, the 0.5–10 keV luminosity during that observation was 2×10^{36} ergs s^{-1}. They could detect the pulsations during this observation with a pulsed fraction of 5.8% in the 0.3–10 keV band (increasing with energy) and a nearly sinusoidal pulse profile. Furthermore, using the same data Kirsch et al. (2004; see also Kirsch & Kendziorra 2003) reported on the mass function of this system and found a minimal mass for the companion star of 0.007 $M_\odot$ when assuming a canonical neutron star mass of 1.4 $M_\odot$. The spectral data are well fit by a continuum model, assumed to be an absorbed Comptonisation model plus a soft component. The latter component only contributed 13% to the flux. Again no emission or absorption lines were found. No detections of the counterparts of the system at other wavelengths have been reported so far.

6. XTE J1814–338

The fifth system (XTE J1814–338) was discovered on June 5, 2003 and has a pulse frequency of 314 Hz (Markwardt et al. 2003b; Fig. 2), with an orbital period of 4.3 hr and a minimum companion mass of 0.15 solar mass (Markwardt et al. 2003d). This 4.3 hr orbital period makes it the widest binary system among the accretion-driven millisecond pulsars and also the one most similar to the general population of low-luminosity neutron star LMXBs (the atoll sources). XTE J1814–338 exhibited many type-I X-ray bursts, which showed burst oscillations with a frequency consistent with the neutron star spin frequency (Markwardt et al. 2003d, Strohmayer et al. 2003). A distance of ~8 kpc was obtained from the only burst which likely reached the Eddington luminosity. The burst oscillations are strongly frequency- and phase-locked to the persistent pulsations (as was also seen for SAX J1808.4–3658; Chakrabarty et al. 2003) and two bursts showed evidence of a frequency decrease of a few tenths of a Hz during the onset of the burst, suggesting a spin down. Strohmayer et al. (2003) also reported on the detection of the first harmonic of the burst oscillations, which is the first time that this has been seen in any burst-oscillation source. This harmonic could arise from two hot-spots on the surface, but Strohmayer et al. (2003) suggested that if the burst oscillations arise from a single bright region, the strength of the harmonic would suggest that the burst emission is beamed (possibly due to a stronger magnetic field strength than in non-pulsating LMXBs). Bhattacharyya et al. (2004) used the non-sinusoidal burst oscillation light curves to constrain the parameters of the neutron star in XTE J1814–338; they obtained a dimensionless radius to mass ratio of $Rc^2/GM > 4.2$ for the neutron star in this source. Their study also suggest that the secondary companion is a hydrogen main sequence star that is significantly bloated (possibly due to X-ray heating).

Wijnands & Homan (2003) analyzed the *RXTE*/PCA data of the source obtained between June 8 and 11, 2003. The overall shape of the 3–60 keV power spectrum is dominated by a strong broad band-limited noise component (Fig. 7), which could be fitted by a broken power-law model with a broad bump superimposed on it at frequencies above the break frequency. These characteristics make the power spectrum of XTE J1814–338 very similar to that observed in the non-pulsing low-luminosity neutron-star LMXBs (the atoll sources) when they are observed at relatively low X-ray luminosities (i.e., in the so-called island state). This is consistent with the hard power-law X-ray spectrum of the source reported by Markwardt et al. (2003d). This resemblance of XTE J1814–338 to the atoll sources was further strengthened (Wijnands & Homan 2003) by the fact that the source is consistent with the relation between the break frequency and the frequency of the bump found for atoll sources by Wijnands & van der Klis (1999). Van Straaten et al. (2005) performed an in-depth analysis of all publicly available *RXTE*/PCA data of XTE J1814–338 to study the power-spectral components and the correlations between their characteristic frequencies. Using those correlations and by comparing them to other sources, they could identify several components that are related to the kHz QPOs. They also found that the frequency correlations were identical to the non-pulsating sources with no need for a frequency shift. This is similar to what they found for XTE J1751–305 but different from SAX J1808.4–3658 and XTE J0929–314 (and for XTE J1807–294 as found by Linares et al. 2005). The reason(s) for this difference between accreting millisecond X-ray pulsars is not know (see Van Straaten et al. 2005 for an extended discussion). From the correlations between the spectral and timing variability it was confirmed that the behavior of XTE J1814–338 was consistent with that of an atoll sources (van Straaten et al. 2005; see also Wijnands & Homan 2003).

Wijnands & Reynolds (2003) reported that the position of XTE J1814–338 was consistent with the *EXOSAT* slew source EXMS B1810–337 which was detected on September 2nd, 1984. If this identification is correct, then its recurrence time can be inferred to be less than 19 years but more than 5 years (the time since the *RXTE*/PCA bulge scan observations started in February 1999), unless the recurrence time of the source varies significantly. Krauss et al. (2003) reported the best position of the source based on a *Chandra* observation and on the detection of the likely optical counterpart of the source (with magnitudes of B = 17.3 and R = 18.8 on June 6). Steeghs (2003) reported on optical spectroscopy of this possible counterpart, specifically on the discovery of prominent hydrogen and helium emission lines, confirming the connection between the optical source and XTE J1814–338.

7. IGR J00291+5934

Very recently, on December 2, 2004, the European Gamma-ray satellite *INTEGRAL* discovered a new X-ray transient named IGR J00291+5934 (Eckert et al. 2004). A day later, *RXTE* observed the source and it was found that this source harbors a 598.88 Hz accretion-driven millisecond X-ray pulsar (Markwardt et al. 2004a). The pulsed amplitude was approximately 6% with no evidence for harmonics. The X-ray spectrum could be fitted with an absorbed power-law model with photon index of 1.7 and a column density of 7×10^{21} cm^{-2}. In Figure 7 the power spectrum between 0.001 and 10,000 Hz is show of the source, clearly showing significant aperiodic variability (see also Markwardt et al. 2004a). Inter-

estingly, of all six accreting millisecond X-ray pulsars the break in the power spectrum is
at the lowest frequency for IGR J00291+5934: during the peak of the outburst the break
frequency for this source was $\sim$0.01 Hz compared to >0.1 Hz for the other sources at their
outburst peaks. Markwardt et al. (2004b) used additional *RXTE* observations to determine
that the orbital period of the system was $\sim$2.45 hours and they obtained a mass function
of $(2.81 \pm 0.02) \times 10^{-5}$ $M_{\odot}$. For a neutron star mass of 1.4 $M_{\odot}$ this results in a lower
limit on the mass for the companion star of 0.038 $M_{\odot}$. The source reached its peak flux (29
mCrab; 2–10 keV) on December 3, 2004, after which the fluxes decreased in a linear way,
until around December 11, 2004, when the rate of decline had increased slightly (Swank &
Markwardt 2004). On December 14, 2004, *Chandra* performed a brief ($\sim$18 ksec) observation of the source using the ACIS-S/HETG combination. Nowak et al. (2004) reported
that the source was at a flux level of $\sim$1 mCrab and its X-ray spectrum could be well-fitted
by an absorbed power-law model with a column density of $\sim 3 \times 10^{21}$ cm^{-2} and a photon
index of $\sim$1.9. A possible iron line feature was also reported by the same authors. The
flux value confirmed the steady decline as seen by Swank & Markwardt (2004) and soon
after this the source could not be detected anymore with *RXTE*/PCA (C. Markwardt 2004,
private communication).

An optical counterpart was proposed by Fox & Kulkarni (2004; see this reference for
a finding chart) with an R magnitude during outburst of $\sim$17.4. The detection of broad
emission lines of HeII and H$_{\alpha}$ from this tentative optical counterpart strongly support this
identification (Roelofs et al. 2004; see also Filippenko et al. 2004). Pooley (2004) found a
1.1 mJy radio source (15 GHz) at a position consistent with that of the optical counterpart
which likely is the radio counterpart of the source which faded during the outburst (Fender
et al. 2004, who also detected the source at 5 GHz at a flux of $\sim$250 μJy), although the
decay did not seem to be very rapid (Rupen et al. 2004). Steeghs et al. (2004) reported the
detection of a decaying infrared counterpart of the source.

After the discovery of this new source, Remillard (2004) constructed a mission-long
light curve using the *RXTE*/ASM data and found that in November 26–28, 1998, and in
September 11–21, 2001, the source might have exhibited other outbursts (see Fig. 11 for the
RXTE/ASM light curve of the December 2004 and September 2001 outbursts of the source).
If confirmed this would give a recurrence time for the outburst of approximately 3 years.
In 't Zand & Heise (2004) did not detected the source with the WFCs aboard *BeppoSAX*
during a net exposure time of 2.9 Msec on the source. The *RXTE*/ASM detections reported
by Remillard (2004) were not covered; the first WFCs data were obtained 16 days and 11
days after the two possible outbursts, respectively, and it is well possible that the source had
by then decayed below the sensitivity limit of the WFCs.

8. Theoretical Work

This chapter is intended to be an observational overview, thus I will not go into detail on the
theoretical papers published on accretion-driven millisecond pulsars. Instead, I will briefly
list some of those papers, most of which focus on SAX J1808.4–3658 since the other five
systems have only been found recently (with IGR J00291+5934 discovered very recently).
Since the discovery of SAX J1808.4–3658, several studies have tried to constrain the properties (i.e., radius, mass, magnetic field strength) of the neutron star in this system (Burderi

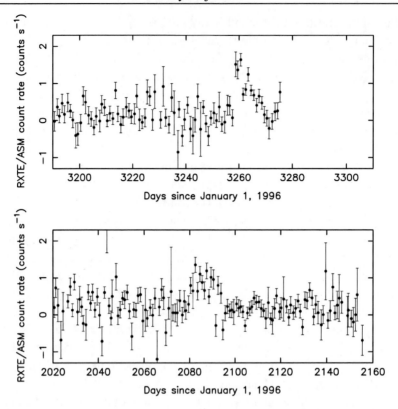

Figure 11. The *RXTE*/ASM light curve of IGR J00291+5934 during the most recent (December 2004) outburst (top panel; near day 3260) and its likely September 2001 outburst (bottom panel; near day 2080). These light curves were made using the public ASM data available at http://xte.mit.edu/ASM_lc.html. The count rates are for the 2–12 keV energy range and are daily averages.

& King 1998; Psaltis & Chakrabarty 1999; Bhattacharyya 2001), while others proposed that the compact object is not a neutron star at all, but instead a strange star (see, e.g., Li et al. 1999; Datta et al. 2000; Zdunik et al. 2000 and references there-in). Other studies focused on the evolutionary history of SAX J1808.4–3658 (Ergma & Antipova 1999) or on the nature of the companion star (Bildsten & Chakrabarty 2001, who suggested a brown dwarf companion star). Recently, Nelson & Rappaport (2003) investigated the evolutionary history of ultracompact binaries such as XTE J0929–314 and XTE J1751–305 (they focused on those two systems but likely their conclusions can also be applied to XTE J1807–294). Rappaport et al. (2004) investigated how accretion could occur in millisecond X-ray pulsars and they postulated that those systems can continue to accrete from a thin disk, even for accretion rates that place the magnetospheric radius well beyond the co-rotation radius.

The discovery of the accretion-driven millisecond pulsars raises an important question: why are those systems different from other neutron star LMXBs for which no pulsations have been found. Cumming et al. (2001; see also Rai Choudhuri & Konar 2002; Cumming 2004) suggested that the low time-averaged accretion rate of SAX J1808.4–3658 might explain why this source is a pulsar. Although the remaining four pulsars were not know at the time of writing of that paper, the same arguments can be used for those systems: when the time-averaged accretion rate is sufficiently high, the neutron star magnetic field might be buried by the accreted matter and does not have time to dissipate through the accreted material. However, for the millisecond X-ray pulsars the time-averaged accretion rate is sufficiently low that the magnetic field dissipation can indeed happen, giving those systems a magnetic field still strong enough to disturb the flow of the accreted matter. However, more neutron-star LMXBs with low time-averaged accretion rate must be found and studied in detail to verify that they all indeed harbor a millisecond pulsar. If an exception is found, the screening model might only be part of the explanation and alternative ideas need to be explored (see, e.g., Titarchuk et al. 2002).

9. Conclusion

From this review it is clear that *RXTE* has played a vital role in the discovery and study of accretion-driven millisecond X-ray pulsars. The detailed studies performed with *RXTE* for those systems have yielded break-throughs in our understanding of kHz QPOs and burst oscillations. Furthermore, three of these accreting pulsars are in ultrashort binaries which will constrain evolutionary paths for this type of systems (e.g., see Nelson & Rappaport 2003). However, it is also clear that the six systems do not form a homogeneous group; their pulsation frequencies span the range between 185 Hz and 599 Hz (Fig. 2), their orbital periods fall between 40 minutes and 4.3 hours, their X-ray light curves are very different (Fig. 5), and also their aperiodic variability properties (Fig. 7). More, well studied outbursts of the currently known systems are needed as well as discoveries of additional systems. At the moment only *RXTE* is capable of performing the necessary timing observations. After *RXTE*, an instrument with similar or better capabilities is highly desirable for our understanding of accretion-driven millisecond pulsars and their connection with the non-pulsating neutron-star LMXBs.

Acknowledgments

I thank Stefanie Wachter for kindly providing the optical data used in Figure 4.

Note added in proofs: On 2005 June 14, the seventh accreting millisecond pulsar (HETE J1900.1–2455) was discovered which has a pulse frequency of 377.3 Hz and an orbital period of 83.3 minutes (Kaaret et al. 2006). The companion star likely has a mass between 0.016 and 0.07 $M_\odot$. Also a single kHz QPO with a frequency of 883 Hz was discovered. Interestingly, not always the pulsations could be detected and the source was still accreting in July 2006, more than 1 year after the source was discovered. Also, a fifth outburst was discovered for SAX J1808.4–3658 during which again kHz QPOs and burst oscillations were observed (Markwardt et al. 2005; Wijnands et al. 2005).

References

Bhattacharya, D., & van den Heuvel, E.P.J., 1991, *Physics Reports*, **203**, 1–124

Bhattacharyya, S., 2001, *ApJ*, **554**, L185–L188

Bhattacharyya, S. et al., 2004, *ApJ*, **619**, 483–491

Bildsten, L. & Chakrabarty, D., 2001, *ApJ*, **557**, 292–296

Burderi, L. & King. A.R., 1998, *ApJ*, **505**, L135–L137

Burderi, L. et al., 2003, *A&A*, **404**, L43–L46

Burgay, M., et al., 2003, *ApJ*, **589**, 902–910

Cacella, P., 2002, *IAUC*, **7893**

Campana, S. et al., 2002, *ApJ*, **575**, L15–L19

Campana, S. et al., 2003, *ApJ*, **594**, L39–L42

Campana, S. et al., 2004, *ApJ*, **614**, L49–L52

Castro-Tirado, A.J. et al., 2002, *IAUC*, **7895**

Chakrabarty, D. 2004, In *Binary Radio Pulsars*, eds. F. A. Rasio & I. H. Stairs, ASP Conf. Ser., **328**, 279

Chakrabarty, D. & Morgan, E. H., 1998, *Nature*, **394**, 346–348

Chakrabarty, D. et al., 2003, *Nature*, **424**, 42–44

Cui, W. et al., 1998, *ApJ*, **504**, L27–L30

Cumming, A. 2004, In *Binary Radio Pulsars* eds. F. A. Rasio & I. H. Stairs, ASP Conf. Ser., **328**, 311

Cumming, A. et al., 2001, *ApJ*, **557**, 958–966

Datta, B. et al., 2000, *A&A*, **355**, L19–L22

Di Salvo, T. & Burderi, L., 2003, *A&A*, **397**, 723–727

Dotani, T. et al., 2000, *ApJ*, **543**, L145–L148

Eckert, D. et al., 2004, *ATEL*, **352**

Ergma, E. & Antipova, J., 1999, *A&A*, **343**, L45–L48

Fender, R. et al., 2004, *ATEL*, **361**

Filippenko, A. V. et al., 2004, *ATEL*, **366**

Ford, E.C., 2000, *ApJ*, **535**, L119–L122

Fox, D. B. & Kulkarni S. R., 2004, *ATEL*, **354**

Gaensler, B.M. et al., 1999, *ApJ*, **522**, L117–L119

Galloway, D.K. et al., 2002, *ApJ*, **576**, L137–L140

Gierlinski, M. & Poutanen, J., 2005, *MNRAS*, **359**, 1261–1276

Gierlinski, M. et al., 2002, *MNRAS*, **331**, 141–153

Giles, A.B. et al., 1998, *IAUC*, **6886**

Giles, A.B. et al., 1999, *MNRAS*, **304**, 47–51

Gilfanov, M. et al., 1998, *A&A*, **338**, L83–L86

Greenhill, J.G. et al., 2002, *IAUC*, **7889**

Heindl, W.A. & Smith, D.M., 1998, *ApJ*, **506**, L35–L38

Heise, J. et al., 1999, *ApL&C*, **38**, 297–300

Homer, L. et al., 2001, *MNRAS*, **325**, 1471–1476

Jonker, P.G. et al., 2000, *ApJ*, **540**, L29–L32

Jonker, P.G. et al., 2003, *MNRAS*, **344**, 201–206

Jonker, P.G. et al., 2004a, *MNRAS*, **349**, 94–98

Jonker, P.G. et al., 2004b, *MNRAS*, **354**, 666–674

Juett, A.M. et al., 2003, *ApJ*, **587**, 754–760

In 't Zand, J. & Heise, J., 2004, *ATEL*, **362**

In 't Zand, J.J.M. et al., 1998, *A&A*, **331**, L25–L28

In 't Zand, J.J.M. et al., 2001, *A&A*, **372**, 916–921

In 't Zand, J.J.M. et al., 2003, *A&A*, **409**, 659–663

Kaaret, P. et al. 2006, *ApJ*, **638**, 963–967

Kato, S., 2004, *PASJ*, **56**, 905–922

Kazarovets, E.V. et al., 2000, *IBVS*, **4870**

Kirsch, M.G.F. & Kendziorra, E., 2003, *ATEL*, **148**

Kirsch, M. G. F. et al., 2004, *A&A*, **423**, L9–L12

Kluzniak, W. et al., 2004, *ApJ*, **603**, L89–L92

Krauss, M.I. et al., 2003, *IAUC*, **8154**

Lamb, F.K. & Miller, M.C., 2004, *ApJ Letters*, submitted (astro-ph/0308179)

Lee, W. H., et al., 2004, *ApJ*, **603**, L93–L96

Li, X.-D. et al., 1999, *Physical Review Letters*, **83**, 3776–3779

Linares, M. et al., 2005, *ApJ*, **634**, 1250–1260

Markwardt, C.B. et al., 2002a, *ApJ*, **575**, L21–L24

Markwardt, C.B. et al., 2002b, *IAUC*, **7993**

Markwardt, C.B. et al., 2003a, *IAUC*, **8080**

Markwardt, C.B. et al., 2003b, *IAUC*, **8144**

Markwardt, C.B. et al., 2003c, *ATEL*, **127**

Markwardt, C.B. et al., 2003d, *ATEL*, **164**

Markwardt, C.B. et al., 2004a, *ATEL*, **353**

Markwardt, C.B. et al., 2004b, *ATEL*, **360**

Markwardt, C.B. et al., 2005, *ATEL*, **505**

Marshall, F.E., 1998, *IAUC*, **6876**

Miller, J.M. et al., 2003, *ApJ*, **583**, L99–L102

Monard, B., 2002, *VSNet alert*, **7550**

Nelson, L. A. & Rappaport, S., 2003, *ApJ*, **598**, 431–445

Nowak, M. A. et al., 2004, *ATEL*, **369**

Pooley, G., 2004, *ATEL*, **355**

Poutanen, J. & Gierlinski, M., 2003, *MNRAS*, **343**, 1301–1311

Psaltis, D. & Chakrabarty, D., 1999, *ApJ*, **521**, 332–340

Rai Choudhuri, A. & Konar, S., 2002, *MNRAS*, **332**, 933–944

Rappaport, S. A. et al., 2004, *ApJ*, **606**, 436–443

Remillard, R.A., 2002, *IAUC*, **7888**

Remillard, R.A. et al., 2002, *IAUC*, **7893**

Remillard, R., 2004, *ATEL*, **357**

Revnivtsev, M.G., 2003, *AstL*, **29**, 383–386

Roche, P. et al., 1998, *IAUC*, **6885**

Roelofs, G. et al., 2004, *ATEL*, **356**

Rupen, M.P. et al., 2002a, *IAUC*, **7893**

Rupen, M.P. et al., 2002b, *IAUC*, **7997**

Rupen, M. et al., 2004, *ATEL*, **364**

Steeghs, D., 2003, *IAUC*, **8155**

Steeghs, D. et al., 2004, *ATEL*, **363**

Stella, L. et al., 2000, *ApJ*, **537**, L115–L118

Strohmayer, T. & Bildsten, L., 2003, To appear in *Compact Stellar X-ray sources*, W.H.G. Lewin & M. van der Klis (eds.), Cambridge University Press (astro-ph/0301544)

Strohmayer, T.E. et al., 1996, *ApJ*, **469**, L9–L12

Strohmayer, T.E. et al., 2003, *ApJ*, **596**, L67–L70

Swank, J.H. & Markwardt, C.B., 2004, *ATEL*, **365**

Titarchuk, L. et al., 2002, *ApJ*, **576**, L49–L52

Tomsick, J.A. et al., 2004, *ApJ*, **610**, 933–940

Vaughan, B.A. et al., 1994, *ApJ*, **435**, 362–371

Van der Klis, M., 2000, *ARA&A*, **38**, 717–760

Van der Klis, M., 2004, To appear in *Compact Stellar X-ray sources*, W.H.G. Lewin & van der Klis (eds.), Cambridge University Press (astro-ph/0410551)

Van der Klis, M. et al., 1996, *ApJ*, **469**, L1–L4

Van der Klis, M. et al., 2000, *IAUC*, **7358**

Van Straaten, S. et al., 2004, In *The Restless High-Energy Universe*, eds. E.P.J. van den Heuvel, J.J.M. in 't Zand, & R.A.M.J. Wijers (Elsevier), *Nuclear Physics B*, **132**, 664–667

Van Straaten, S. et al., 2005, *ApJ*, **619**, 455–482

Wachter, S. & Hoard, D.W., 2000, *IAUC*, **7363**

Wachter, S. et al., 2000, *HEAD*, **32**, 24.15

Wang, Z. et al., 2001, *ApJ*, **563**, L61–L64

Wijnands, R., 2003, *ApJ*, **588**, 425–429

Wijnands, R., 2004a, In *The Restless High-Energy Universe*, eds. E.P.J. van den Heuvel, J.J.M. in 't Zand, & R.A.M.J. Wijers (Elsevier), *Nuclear Physics B*, **132**, 496–505

Wijnands, R., 2004b, In *X-ray Timing 2003: Rossi and Beyond*, eds. P. Kaaret, F.K. Lamb, & J.H. Swank (Melville, NY), *AIP*, **714**, 209–216

Wijnands, R. & van der Klis, M., 1998a, *Nature*, **394**, 344–346

Wijnands, R. & van der Klis, M., 1998b, *ApJ*, **507**, L63–L66

Wijnands, R. & van der Klis, M., 1999, *ApJ*, **514**, 939–944

Wijnands, R. & Homan, J., 2003, *ATEL*, **165**

Wijnands, R. & Reynolds, A., 2003, *ATEL*, **166**

Wijnands, R. et al., 2001, *ApJ*, **560**, 892–896

Wijnands, R. et al., 2002, *ApJ*, **571**, 429–434

Wijnands, R. et al., 2003, *Nature*, **424**, 44–47

Wijnands, R. et al., 2005, *ApJ*, **619**, 492–502

Wijnands, R. et al., 2005, *ATEL*, **507**

Zdunik, J.L. et al., 2000, *A&A*, **359**, 143–147

INDEX

S

T

U